Alan Chan
Oct '76
Berkeley, Ca.

Studies in Dynamical Systems

During the last twenty years there has been a progressive increase in the complexity and degree of interconnection of systems of all kinds. The reasons are clear: recent progress in communication, data processing, and control have made possible a much greater degree of coordination between the parts of a system than was ever possible before. Digital computers have played a central part in this evolution, and are likely to play a still greater part in the future.

In the course of this development the traditional engineering disciplines have tended to polarize. At one extreme is the detailed consideration of devices, drawing heavily on applied physics and chemistry. At the other extreme is the consideration of aggregates of devices and their interaction and interrelation. The distinction is not always easy to maintain, because what is called a system in one context may become an element when still larger systems are considered. Nevertheless the two kinds of approach are sufficiently clear and each has its special techniques.

It is the aim of this series to give an account of the theory and practice involved in designing dynamical engineering systems. By this we imply an interest in deterministic and stochastic behaviour, control, optimization, and synthesis. Our effort has been to assemble a series which is oriented towards problem solving in these areas, but also presents a sound conceptual framework with an eye towards generalization.

The level of presentation will depend to some extent on the topic, but will generally cover the range from introductory undergraduate to first-year postgraduate courses. The success of the series must depend ultimately on its acceptance both by students and by practising engineers.

Roger W. Brockett
Harvard University
H. H. Rosenbrock
University of Manchester Institute of Science and Technology

Studies in Dynamical Systems

Editors

Roger W. Brockett, *Harvard University*

H. H. Rosenbrock, *University of Manchester Institute of Science and Technology*

H. H. ROSENBROCK
State-space and Multivariable Theory

H. H. ROSENBROCK and C. STOREY
Mathematics of Dynamical Systems

J. L. WILLEMS
Stability Theory of Dynamical Systems

Further titles in preparation

State-space and Multivariable Theory

H. H. Rosenbrock
University of Manchester Institute of Science and Technology

Wiley Interscience Division
JOHN WILEY & SONS, INC., NEW YORK

Library of Congress catalog number: 75–37331

Made and printed in Great Britain by William Clowes and Sons, Limited
London and Beccles

juvat integros accedere fonteis,
Atque haurire : juvatque novos decerpere flores

Lucretius, DE RERUM NATURA

Contents

6 Time-dependent systems

7 Notes and references

Preface

This book has arisen from efforts over many years to find simple and powerful methods for designing industrial control systems. Experience has shown that the classical transfer-function methods were very suitable for isolated loops, but failed in the multivariable situation. The newer state-space methods at first promised success, but came in the end to seem disappointing.

A growing conviction that design methods could only arise from deeper understanding led to a programme of theoretical research. This had as one aim the development of a theory which would embrace transfer-function and state-space descriptions in a single whole. The hope was that on such a basis could be founded more powerful design techniques allied to both.

This work has now come to a stage where it can be presented as a coherent theory. The simplest situation, of a linear constant system, can now be analysed in some detail. Design methods based on these results have been developed and are now being applied. They will be described in detail elsewhere, but some of the theoretical basis is given here in Chapter 5. Time-varying systems are much less well understood, and are therefore treated only briefly. Nonlinear systems are not mentioned as almost nothing is known about them in the present context.

Besides its practical use, the theory presented gives a much more connected account than was previously available and should therefore appeal to students. Results which were obtained in isolation are now shown to be closely related. The amount of structure possessed by apparently simple systems is truly astonishing, and there is scope for much further research.

My debt to other workers in this field will be obvious, and is partly described in Chapter 7. It is also a pleasure to acknowledge helpful discussions over many years with friends and colleagues. Chief among these have been Professor L. A. Gould, Professor R. W. Brockett, Professor A. G. J. MacFarlane, Professor C. Storey, and Dr. S. Barnett. My thanks are also gratefully given to Dr. N. Munro and to many students who read and commented on the draft; and to Mrs. C. H. McKay who produced an immaculate typescript from a lamentably untidy manuscript. Finally, and more generally, I wish to thank Professor J. F. Coales and the Science Research Council for making possible a late transition from industry to university of which this book is one result.

H.H.R.

1 Preliminary mathematics

This chapter contains a summary of the principal mathematical results which are needed later in the book. These are presented in a way which is intended rather for reference than for consecutive reading. Nevertheless, a brief scanning of the chapter will indicate to the reader any major areas where his mathematical preparation needs to be strengthened.

Almost all the necessary mathematics is contained in Rosenbrock and Storey (1970) which will be referred to throughout this book as R & S. Where proofs are given in R & S, they are not repeated here but the appropriate reference is given. In the few cases where results are not to be found in R & S, proofs are given in this chapter.

1 Matrices and determinants

An *$m \times n$ matrix A over a field F* is an array

$$A = (a_{ij}) = \begin{pmatrix} a_{11} & a_{12} & \dots & a_{1n} \\ a_{21} & a_{22} & \dots & a_{2n} \\ \vdots & \vdots & & \vdots \\ a_{m1} & a_{m2} & \dots & a_{mn} \end{pmatrix} \tag{1.1}$$

where each $a_{ij} \in F$. The *transpose* A^{T} of A is the $n \times m$ matrix having ijth element a_{ji}. The *leading diagonal* of A is the set of elements $(a_{11}, a_{22}, \dots, a_{mm})$ if $m < n$, or $(a_{11}, a_{22}, \dots, a_{nn})$ if $n < m$: when $m = n$ the leading diagonal is called the *principal diagonal*. The sum of two $m \times n$ matrices A, B is the matrix having ijth element $a_{ij} + b_{ij}$,

$$A + B = (a_{ij}) + (b_{ij}) = (a_{ij} + b_{ij}) \tag{1.2}$$

The product of an $m \times n$ matrix A with an $n \times r$ matrix B is the $m \times r$ matrix having ijth element $\sum_{k=1}^{n} a_{ik}b_{kj}$,

$$AB = (a_{ik})(b_{kj}) = \left(\sum_{k=1}^{n} a_{ik}b_{kj}\right) \tag{1.3}$$

If A is $n \times n$, two *scalars* (that is, elements of the field F) associated with A are its *trace* (or *spur*) and its *determinant*. The trace of A is the sum of the elements on its principal diagonal, whence trace $(A + B) =$ trace $A +$ trace B. The determinant det A, or $|A|$, is defined recursively by

$$|A| = (-1)^{i-1}[a_{i1}M_{i1} - a_{i2}M_{i2} + a_{i3}M_{i3} - \cdots (-1)^{n-1}a_{in}M_{in}] \tag{1.4}$$

where M_{ij} is the $(n-1) \times (n-1)$ *minor* of a_{ij}, and is the determinant of the matrix formed from A by striking out row i and column j. The signed minor $(-1)^{i+j}M_{ij}$ is called the cofactor A_{ij} of a_{ij}, so that (1.4) can be written

$$|A| = \sum_{j=1}^{n} a_{ij}A_{ij} \tag{1.5}$$

By (1.4), $|A|$ is made to depend ultimately on determinants of scalars, and these are defined to be equal to the scalars,

$$\det a_{ij} = a_{ij} \tag{1.6}$$

Notice that $|(a_{ij})|$ is a notation for $|A|$, and not for det a_{ij}. Confusion with the modulus $|a_{ij}|$ is avoided by using the bars to denote the determinant only of matrices having dimension $n \geqslant 2$. We have, for A, B both $n \times n$,

$$|A^{\mathrm{T}}| = |A| \tag{1.7}$$

$$|AB| = |A|\,|B| \tag{1.8}$$

$$|aA| = |aI_n|\,|A| = a^n|A| \tag{1.9}$$

where a is a scalar. Interchange of two rows in a matrix reverses the sign of, but otherwise leaves unchanged, its determinant. By (1.7) the same is true for the interchange of two columns. Addition of any scalar multiple of one row (resp. column) to another row (resp. column) in a matrix leaves its determinant unchanged.

In addition to the $(n-1) \times (n-1)$ minors M_{ij}, minors of other orders can be defined. A minor formed from A by striking out all except rows $i_1, i_2, \ldots, i_r$ and all except columns $j_1, j_2, \ldots, j_r$, and then taking the determinant of the resulting $r \times r$ array, is called a *minor of order r*, and is written

$$A^{i_1, i_2, \ldots, i_r}_{j_1, j_2, \ldots, j_r} \tag{1.10}$$

With this notation

$$A_{ij} = (-1)^{i+j} A^{1, 2, \ldots, i-1, i+1, \ldots, n}_{1, 2, \ldots, j-1, j+1, \ldots, n} \tag{1.11}$$

The *leading minors* of A are $A^{1}_{1} = a_{11}$, $A^{1,2}_{1,2}, \ldots, A^{1,2,\ldots,n}_{1,2,\ldots,n} = |A|$.

An $n \times n$ matrix A over F is *singular* if $|A| = 0$, the zero of F. If A is non-singular it has an *inverse* A^{-1}, which has as its ijth element $A_{ji}/|A|$. The matrix having ijth element A_{ji} is called the *adjoint* of A, adj A. Hence

$$A^{-1} = (\operatorname{adj} A)\,|A|^{-1} \tag{1.12}$$

The product of A and its inverse is the *identity matrix I*,

$$AA^{-1} = A^{-1}A = I \tag{1.13}$$

where

$$I = (\delta_{ij}), \quad \delta_{ij} = \begin{cases} 0, & i \neq j \\ 1, & i = j \end{cases} \tag{1.14}$$

and δ_{ij} is called the *Kronecker delta.*

If the elements a_{ij} belong to a ring R, rather than a field, what has been said remains true (after the appropriate changes) with one exception: for the existence of A^{-1} as a matrix over R it is no longer sufficient to have $|A| \neq 0$. The determinant is now an element of R, and if (1.12) is to have meaning, $|A|$ must have an inverse. In other words, $|A|$ must be *regular* in R (that is, must be a *unit* in R).

As an example the polynomials with coefficients in F form a ring R, which has as its regular elements the non-zero elements of F. Only those polynomial matrices having as determinant a non-zero element of F have an inverse which is also a polynomial matrix. Such matrices are occasionally called *regular*, but the term *unimodular* is preferred here as 'regular' has another meaning for polynomial matrices: a polynomial matrix is called 'regular' when it can be written

$$A(s) = A_0 + A_1 s + \cdots + A_r s^r \tag{1.15}$$

and A_r is non-singular.

If we regard the polynomials in s as functions of s, a matrix A over polynomials is also a function of s, and we write $A(s)$ or $A(\cdot)$ (compare Section 11). In special circumstances $A(s)$ may be the same for all s, and we express this by saying that A is *constant.* A similar terminology is used when the elements of A are functions (no longer necessarily polynomials) of a variable t.

Theorem 1.1 (*Laplace's expansion* of a determinant) Let A be an $n \times n$ matrix over a ring R (or, as a special case, over a field F). Then

$$|A| = \sum_{j} (-1)^{i_1 + i_2 + \cdots + i_r + j_1 + j_2 + \cdots + j_r} A^{i_1, i_2, \ldots, i_r}_{j_1, j_2, \ldots, j_r} A^{i'_1, i'_2, \ldots, i'_{n-r}}_{j'_1, j'_2, \ldots, j'_{n-r}} \tag{1.16}$$

where $i_1 < i_2 < \cdots < i_r$ and $i'_1 < i'_2 < \cdots < i'_{n-r}$ form a complete set of indices $1, 2, \ldots, n$; $j_1 < j_2 < \cdots < j_r$ and $j'_1 < j'_2 < \cdots < j'_{n-r}$ form another such set; and the summation in (1.16) extends over all choices of the r indices $j_1, j_2, \ldots, j_r$.

Proof Equation (1.16) is given in R & S, p. 41, but with

$$\delta^{i_1, i_2, \ldots, i_r, i'_1, i'_2, \ldots, i'_{n-r}}_{j_1, j_2, \ldots, j_r, j'_1, j'_2, \ldots, j'_{n-r}} \tag{1.17}$$

replacing

$$(-1)^{i_1 + i_2 + \cdots + i_r + j_1 + j_2 + \cdots + j_r} \tag{1.18}$$

The symbol (1.17) is $+1$ if $i_1, i_2, \ldots, i_r, i'_1, i'_2, \ldots, i'_{n-r}$ is an even permutation of $j_1, j_2, \ldots, j_r, j'_1, j'_2, \ldots, j'_{n-r}$, and is -1 otherwise. Now the first of these sequences can be transformed to the second by way of a preliminary transformation into the sequence $1, 2, \ldots, n$. In transforming $1, 2, \ldots, n$ into $i_1, i_2, \ldots, i_r, i'_1, i'_2, \ldots, i'_{n-r}$, the number of reversals of order (which is the same as in the reverse transformation) is

$$(i_1 - 1) + (i_2 - 2) + \cdots + (i_r - r) \tag{1.19}$$

because $i_1 < i_2 < \cdots < i_r, i'_1 < i'_2 < \cdots < i'_{n-r}$, and once the $i_1, i_2, \ldots, i_r$ have been reordered the $i'_1, i'_2, \ldots, i'_{n-r}$ are consequently also reordered. Similarly in transforming $1, 2, \ldots, n$ into $j_1, j_2, \ldots, j_r, j'_1, j'_2, \ldots, j'_{n-r}$ the number of reversals of order is

$$(j_1 - 1) + (j_2 - 2) + \cdots + (j_r - r) \tag{1.20}$$

The total number of reversals of order in the permutation corresponding to (1.17) is therefore

$$i_1 + i_2 + \cdots + i_r + j_1 + j_2 + \cdots + j_r - r(r+1) \tag{1.21}$$

Consequently (1.18) gives the same sign as (1.17).

A matrix is said to be *partitioned* when it is written in terms of submatrices, for example

$$P = \begin{pmatrix} T & U \\ -V & W \end{pmatrix} \tag{1.22}$$

where P is $(r + m) \times (r + l)$, T is $r \times r$, U is $r \times l$, V is $m \times r$, W is $m \times l$. The submatrices T, U, etc. are called *blocks*. It readily follows from Laplace's expansion that when $l = m$

$$\begin{vmatrix} T & 0 \\ -V & W \end{vmatrix} = \begin{vmatrix} T & U \\ 0 & W \end{vmatrix} = |T|\,|W| \tag{1.23}$$

Theorem 1.2 If T, U, V, W in (1.22) are matrices over a field F (for example, over the field of rational functions) and if $l = m$ and $|T|$ is not zero in F, then

$$|P| = |T|\,|VT^{-1}U + W| \tag{1.24}$$

Proof We may write

$$P = \begin{pmatrix} T & 0 \\ 0 & I_m \end{pmatrix} \begin{pmatrix} I_r & 0 \\ -V & I_m \end{pmatrix} \begin{pmatrix} I_r & T^{-1}U \\ 0 & VT^{-1}U + W \end{pmatrix} \tag{1.25}$$

and the result follows as in (1.23).

Theorem 1.3 Let A, B be matrices over a ring R, respectively $l \times n$, $n \times m$. Then the $r \times r$ minors of the product matrix $C = AB$ are

$$C^{i_1, i_2, \ldots, i_r}_{j_1, j_2, \ldots, j_r} = \sum_k A^{i_1, i_2, \ldots, i_r}_{k_1, k_2, \ldots, k_r} B^{k_1, k_2, \ldots, k_r}_{j_1, j_2, \ldots, j_r} \tag{1.26}$$

where

$$1 \leqslant i_1 < i_2 < \cdots < i_r \leqslant l \tag{1.27}$$

$$1 \leqslant j_1 < j_2 < \cdots < j_r \leqslant m \tag{1.28}$$

and the sum is taken over all indices k_q satisfying

$$1 \leqslant k_1 < k_2 < \cdots < k_r \leqslant n \tag{1.29}$$

Proof Let L be the $r \times n$ matrix consisting of rows $i_1, i_2, \ldots, i_r$ of A, and let M be the $n \times r$ matrix consisting of columns $j_1, j_2, \ldots, j_r$ of B. Then from (1.3) and (1.24) it follows that

$$C^{i_1, i_2, \ldots, i_r}_{j_1, j_2, \ldots, j_r} = |LM| = \begin{vmatrix} I_n & M \\ -L & 0 \end{vmatrix} \tag{1.30}$$

Now carry out a Laplace expansion of the last determinant in (1.30) in terms of the first n rows and last r rows. Equation (1.16) gives, on omitting terms which are zero,

$$\begin{vmatrix} I_n & M \\ -L & 0 \end{vmatrix} = \sum_k (-1)^{n+1+n+2+\cdots+n+r+k_1+k_2+\cdots+k_r}(-1)^r \times L^{1,2,\ldots,r}_{k_1,k_2,\ldots,k_r}(I_n, M)^{1,2,\ldots,n}_{k'_1,k'_2,\ldots,k'_n} \tag{1.31}$$

where $1 \leqslant k_1 < k_2 < \cdots < k_r \leqslant n$ and $k'_1 < k'_2 < \cdots < k'_n$ form a complete set of indices $1, 2, \ldots, n + r$ and the summation is over all choices of k_1, $k_2, \ldots, k_r$. But it is easy to see by interchanging rows that

$$(I_n, M)^{1,2,\ldots,n}_{k'_1,k'_2,\ldots,k'_n} = (-1)^{n-k_r+n-1-k_{r-1}+\cdots+n-r+1-k_1} M^{k_1,k_2,\ldots,k_r}_{1,2,\ldots,r} \tag{1.32}$$

It follows that the sign in (1.31) is positive, and because

$$L^{1,2,\ldots,r}_{k_1,k_2,\ldots,k_r} = A^{i_1,i_2,\ldots,i_r}_{k_1,k_2,\ldots,k_r} \tag{1.33}$$

$$M^{k_1,k_2,\ldots,k_r}_{1,2,\ldots,r} = B^{k_1,k_2,\ldots,k_r}_{j_1,j_2,\ldots,j_r} \tag{1.34}$$

this proves (1.26).

Theorem 1.4 Let A be an $n \times n$ matrix over a field F with $|A|$ non-zero in F. Then the $r \times r$ minors of $B = A^{-1}$ are

$$B^{i_1, i_2, \ldots, i_r}_{j_1, j_2, \ldots, j_r} = (-1)^{i_1+i_2+\cdots+i_r+j_1+j_2+\cdots+j_r} A^{j'_1, j'_2, \ldots, j'_{n-r}}_{i'_1, i'_2, \ldots, i'_{n-r}} |A|^{-1} \tag{1.35}$$

where $i_1 < i_2 < \cdots < i_r$, and $i'_1 < i'_2 < \cdots < i'_{n-r}$ form a complete set of indices $1, 2, \ldots, n$; and $j_1 < j_2 < \cdots < j_r$ and $j'_1 < j'_2 < \cdots < j'_{n-r}$ form another such set.

Proof On applying (1.26) to

$$AB = I \tag{1.36}$$

We obtain

$$\sum_i A^{k_1, k_2, \ldots, k_r}_{i_1, i_2, \ldots, i_r} B^{i_1, i_2, \ldots, i_r}_{j_1, j_2, \ldots, j_r} = \begin{cases} 1, & \sum_{p=1}^{r} (k_p - j_p)^2 = 0 \\ 0, & \sum_{p=1}^{r} (k_p - j_p)^2 > 0 \end{cases} \tag{1.37}$$

On the other hand equation (1.16) gives

$$\sum_i A^{k_1, k_2, \ldots, k_r}_{i_1, i_2, \ldots, i_r} (-1)^{j_1 + j_2 + \cdots + j_r + i_1 + i_2 + \cdots + i_r} A^{j'_1, j'_2, \ldots, j'_{n-r}}_{i'_1, i'_2, \ldots, i'_{n-r}}$$

$$= \begin{cases} |A|, & \sum_{p=1}^{r} (k_p - j_p)^2 = 0 \\ 0, & \sum_{p=1}^{r} (k_p - j_p)^2 > 0 \end{cases} \tag{1.38}$$

where $i_1 < i_2 < \cdots < i_r$ and $i'_1 < i'_2 < \cdots < i'_{n-r}$ form a complete set of indices $1, 2, \ldots, n$; $j_1 < j_2 < \cdots < j_r$ and $j'_1 < j'_2 < \cdots < j'_{n-r}$ form another such set; and we have used the fact that a matrix having two identical rows has zero determinant. Now there are $N = \binom{n}{r}$ possible combinations $i_1, i_2, \ldots, i_r$ and similarly for $j_1, j_2, \ldots, j_r$ and $k_1, k_2, \ldots, k_r$. Let these be arranged in lexicographic order, and assemble the $A^{k_1, k_2, \ldots, k_r}_{i_1, i_2, \ldots, i_r}$ and $B^{i_1, i_2, \ldots, i_r}_{j_1, j_2, \ldots, j_r}$ into $N \times N$ matrices according to this ordering of rows and columns. Call the resulting matrices $\mathscr{A}$ and $\mathscr{B}$. Then equation (1.37) is just

$$\mathscr{A}\mathscr{B} = I_N \tag{1.39}$$

The solution $\mathscr{B}$ of this equation is unique, and comparison of (1.37) and (1.38) therefore establishes (1.35).

The *determinantal rank* of a matrix over a ring R is r if there is a minor of order r which is not zero in R, whereas every minor of order $r + 1$ is zero in R. The *rank defect* of an $m \times n$ matrix over R having rank r is the smaller of $m - r$, $n - r$. Theorem 1.3 shows that if rank $A = r_a$, rank $B = r_b$, then rank $C \leqslant$ min (r_a, r_b). Moreover, if A is $p \times n$ with rank $n \leqslant p$, B is $n \times q$ with rank $n \leq q$, then rank AB is n. For by Theorem 1.3, the $n \times n$ minors of AB consist of all products of an $n \times n$ minor of A with an $n \times n$ minor of B, while all $(n+1) \times (n+1)$ minors of AB must be zero.

A square matrix A is *upper triangular* if $a_{ij} = 0$ when $i > j$. It is *lower triangular* if $a_{ij} = 0, j > i$. A *triangular* matrix is one which is either upper or lower triangular. A *diagonal matrix* is one which is both upper triangular and lower triangular. A *permutation matrix* has all entries zero except for a single entry 1 in each column, no two such entries being in the same row.

1.1 Equivalence

The *elementary row operations* on a matrix over a field F are defined in the following way:

(i) Multiply any row by a non-zero element $\alpha \in F$.
(ii) Add to any row a multiple by any $\beta \in F$ of any other row.
(iii) Interchange any two rows.

Elementary *column* operations are defined in a similar way.

Theorem 1.5 (*Gauss reduction*) By elementary row operations (and possibly by reordering the columns) any $m \times n$ matrix over F can be reduced to the form

$$\begin{pmatrix} I_r & Q \\ 0_{m-r,r} & 0_{m-r,n-r} \end{pmatrix} \tag{1.40}$$

where Q is an $r \times (n-r)$ submatrix, not necessarily zero. A similar theorem for column operations is obtained by transposition.

Proof See R & S, p. 48.

The elementary row operations for a polynomial matrix are defined in the following way:

(i) Multiply any row by a non-zero element $\alpha \in F$, where F is the field of the coefficients of the polynomials.
(ii) Add to any row a multiple by any polynomial of any other row.
(iii) Interchange any two rows.

Elementary column operations are defined in a similar way. Such operations are all represented by unimodular matrices, and any unimodular matrix can be obtained as a product of elementary operations (row, or column, or both as desired).

Theorem 1.6 By elementary row (resp. column) operations an $n \times n$ polynomial matrix $A(s)$ can be brought to upper (or lower, as desired) triangular form.

Proof In the first column, if all elements are not zero select an element of lowest degree and bring it to position (1, 1) by a row interchange. Subtract appropriate multiples of the first row from each other row to reduce the degree of the $n - 1$ elements in positions (2, 1), (3, 1), . . ., (n, 1) below that of the element in position (1, 1). If not all these $n - 1$ elements are zero, bring an element of lowest degree to position (1, 1). This procedure ends in reducing to zero all elements in the first column except (possibly) the element in position (1, 1): note that this is a common divisor of all elements in the first column of A. Proceed similarly with the $(n - 1) \times (n - 1)$ submatrix which excludes the first column and first row. In this way the matrix is reduced to upper triangular form. Lower triangular form can be obtained in a similar way. Column operations can be used throughout in place of row operations.

Theorem 1.7 (*Smith form*) By a combination of elementary row and column operations, an $m \times n$ polynomial matrix $A(s)$ can be reduced to Smith form

$$S(s) = M(s)A(s)N(s) \tag{1.41}$$

where M, N are unimodular and represent the row and column operations respectively and

$$S(s) = \begin{pmatrix} Q(s) & 0_{m,n-m} \end{pmatrix}, \quad n > m \tag{1.42}$$

$$S(s) = Q(s), \quad n = m \tag{1.43}$$

$$S(s) = \begin{pmatrix} Q(s) \\ 0_{m-n,n} \end{pmatrix}, \quad n < m \tag{1.44}$$

Here $Q(s)$ is a diagonal matrix having as the entries on its principal diagonal the *invariant polynomials* $\varepsilon_i(s)$. Each non-zero invariant polynomial is *monic*; that is, has the coefficient of its highest power of s equal to 1. If the rank of $A(s)$ is r, there are r non-zero invariant polynomials, occupying the leading r positions, and the remaining invariant polynomials are zero. Each $\varepsilon_i(s)$ divides $\varepsilon_{i+1}(s)$, $i = 1, 2, \ldots, r - 1$. The $\varepsilon_i(s)$ are given by

$$\varepsilon_i(s) = \frac{D_i(s)}{D_{i-1}(s)}, \quad i = 1, 2, \ldots, r \tag{1.45}$$

where $D_0(s) = 1$ by definition, and $D_i(s)$ is the greatest common divisor of minors of order i in A, $i = 1, 2, \ldots, n$.

Proof See R & S, pp. 91–93.

Let the invariant polynomials $\varepsilon_i(s)$ be factorized into their monic irreducible factors $\phi_j(s)$ over the field of the coefficients. Let the power of ϕ_j occurring in ε_i be k_{ij}. Then those of $\phi_j^{k_{ij}}$ having $k_{ij} \neq 0$ are called the *elementary divisors* of A.

The reductions in Theorems 1.5, 1.6, and 1.7 can all be written

$$A_1 = MAN \tag{1.46}$$

where M, N are non-singular matrices, and if polynomial are unimodular. This transformation is called *equivalence.*

Theorem 1.8 If the $n \times n$ polynomial matrices $sI_n - A$, $sI_n - A_1$ are equivalent, where A, A_1 are constant, then the transforming matrices may be chosen constant. That is, if

$$sI_n - A_1 = M(s)(sI_n - A)N(s) \tag{1.47}$$

with $M(s)$, $N(s)$ unimodular, then

$$sI_n - A_1 = M_0(sI_n - A)N_0 \tag{1.48}$$

where M_0, N_0 are constant. Moreover, $M_0 = N_0^{-1}$,

$$M(s) = (sI_n - A_1)M_1(s) + M_0 \qquad \text{and} \qquad N(s) = N_1(s)(sI_n - A_1) + N_0$$

where $M_1(s)$, $N_1(s)$ are polynomial matrices.

Proof See R & S, pp. 96, 97.

The $m \times n$ polynomial matrix $sA + B$, where A, B are constant, is called a *pencil of matrices.* It is *regular* if $m = n$ and $|sA + B| \not\equiv 0$: otherwise it is *singular.*

When A, A_1 are $n \times n$ matrices over a field and are related as in (1.48) by

$$A_1 = H^{-1}AH \tag{1.49}$$

then A_1 and A are said to be *similar*. A necessary and sufficient condition for A, A_1 to be similar is that $sI_n - A$, $sI_n - A_1$ have the same invariant polynomials. If

$$A_1 = H^{\mathrm{T}}AH \tag{1.50}$$

then A_1 and A are said to be *congruent.*

Under similarity there are three important standard forms.

(i) *First natural normal form*, namely

$$A_1 = H^{-1}AH = \begin{pmatrix} C_1 & 0 & \dots & 0 \\ 0 & C_2 & \dots & 0 \\ \vdots & \vdots & & \vdots \\ 0 & 0 & \dots & C_p \end{pmatrix} \tag{1.51}$$

Here the C_i are *companion matrices*, having the form

$$\begin{pmatrix} 0 & 0 & \dots & 0 & -c_0^i \\ 1 & 0 & \dots & 0 & -c_1^i \\ 0 & 1 & \dots & 0 & -c_2^i \\ \vdots & \vdots & & \vdots & \vdots \\ 0 & 0 & \dots & 1 & -c_{p_i-1}^i \end{pmatrix} \tag{1.52}$$

To each invariant polynomial $\varepsilon_i(s)$ of $sI_n - A$ corresponds one companion matrix C_i, and

$$\varepsilon_i(s) = |sI - C_i| \tag{1.53}$$

The companion matrices can be given three other forms, obtained by reflection in the two diagonals.

(ii) *Second natural normal form.* This has companion matrices corresponding to the *elementary divisors* of A.

(iii) *Jordan form* (when F is the field of complex numbers), namely

$$A_1 = H^{-1}AH = \begin{pmatrix} J_1 & 0 & \dots & 0 \\ 0 & J_2 & \dots & 0 \\ 0 & 0 & \dots & J_q \end{pmatrix} \tag{1.54}$$

where the *Jordan blocks* J_i are

$$J_i = \begin{pmatrix} \lambda_i & 1 & 0 & \dots & 0 & 0 \\ 0 & \lambda_i & 1 & \dots & 0 & 0 \\ \vdots & \vdots & \vdots & & \vdots & \vdots \\ 0 & 0 & 0 & \dots & \lambda_i & 1 \\ 0 & 0 & 0 & \dots & 0 & \lambda_i \end{pmatrix} \tag{1.55}$$

To each elementary divisor of A corresponds a Jordan block, and

$$|sI - J_i| = (s - \lambda_i)^{q_i} \tag{1.56}$$

If A is real and symmetric, the λ_i are all real, there are n Jordan blocks, and H may be chosen *orthogonal*; that is, such that $H^{\mathrm{T}} = H^{-1}$ (R & S, p. 103).

From (1.54) it follows that

$$|sI_n - A| = \prod_{i=1}^{q} |sI - J_i| = \prod_{i=1}^{q} (s - \lambda_i)^{q_i} \tag{1.57}$$

and the λ_i are therefore the n roots of the polynomial $|sI_n - A|$. These quantities

λ_i are termed the *eigenvalues* of A. Because $\lambda_i I_n - A$ is singular, there is at least one vector v_i (see below, Section 4) such that

$$(\lambda_i I_n - A)v_i = 0 \tag{1.58}$$

and v_i is called an *eigenvector*. Since

$$|\lambda I_n - A^{\mathrm{T}}| = |\lambda I_n - A| \tag{1.59}$$

it follows that the eigenvalues of A^{T} are those of A. Taking complex conjugates,

$$\overline{|\lambda I_n - A|} = |\bar{\lambda} I_n - \bar{A}| \tag{1.60}$$

so that the eigenvalues of $\bar{A}$ are $\bar{\lambda}_i$. If $A^* = \bar{A}^{\mathrm{T}}$, it follows that A^* has eigenvalues $\bar{\lambda}_i$.

Theorem 1.9 (*Gershgorin's theorem*) The eigenvalues of an $n \times n$ matrix A over the complex numbers lie in the union of the disks

$$|z - a_{ii}| \leqslant \sum_{\substack{j=1 \\ j \neq i}}^{n} |a_{ij}|, \quad i = 1, 2, \ldots, n \tag{1.61}$$

and also in the union of the disks

$$|z - a_{ii}| \leqslant \sum_{\substack{j=1 \\ j \neq i}}^{n} |a_{ji}|, \quad i = 1, 2, \ldots, n \tag{1.62}$$

Proof See R & S, pp. 105, 106.

1.2 Linear equations

The equation

$$Ax = b \tag{1.63}$$

represents the set of linear equations

$$\sum_{j=1}^{n} a_{ij}x_j = b_i, \quad i = 1, 2, \ldots, m \tag{1.64}$$

If $m = n$ and A is non-singular, equation (1.63) has the unique solution

$$x = A^{-1}b \tag{1.65}$$

If $m \neq n$, or if $m = n$ and $|A| = 0$, equation (1.63) may have no solution (equation (1.64) is then *inconsistent*) or it may have one solution or multiple solutions (equation (1.64) is then *consistent*).

Theorem 1.10 Equations (1.64) are consistent if and only if the rank of A in equation (1.63) is the same as the rank of the *augmented matrix* B given by

$$B = (A \quad b) = \begin{pmatrix} a_{11} & a_{12} & \dots & a_{1n} & b_1 \\ a_{21} & a_{22} & \dots & a_{2n} & b_2 \\ \vdots & \vdots & & \vdots & \vdots \\ a_{m1} & a_{m2} & \dots & a_{mn} & b_m \end{pmatrix} \tag{1.66}$$

Proof See R & S, pp. 48–50.

2 The characteristic matrix

Some general properties of polynomial matrices were given in Section 1. Here we study the polynomial matrix $sI_n - A$ which is associated with an $n \times n$ matrix A over the real or complex numbers: $sI_n - A$ is called the *characteristic matrix* of A. The determinant of $sI_n - A$ is a monic polynomial Δ, known as the *characteristic polynomial* of A,

$$\Delta(s) = |sI_n - A| = s^n + a_{n-1}s^{n-1} + \cdots + a_1 s + a_0 \tag{2.1}$$

Theorem 2.1 (*Cayley–Hamilton theorem*) Every $n \times n$ matrix A over a field F satisfies its own characteristic equation; that is

$$\Delta(A) = A^n + a_{n-1}A^{n-1} + \cdots + a_1 A + a_0 I_n = 0 \tag{2.2}$$

Proof See R & S, pp. 87–88.

Theorem 2.2 The inverse of $sI_n - A$ (known as the *resolvent matrix*) is given by the formula:

$$\begin{aligned}(sI_n - A)^{-1} = \{&(s^{n-1} + a_{n-1}s^{n-2} + \cdots + a_1)I_n \\ &+ (s^{n-2} + a_{n-1}s^{n-3} + \cdots + a_2)A \\ &+ \cdots + (s + a_{n-1})A^{n-2} + A^{n-1}\} \div \Delta(s)\end{aligned} \tag{2.3}$$

Proof Multiply the expression on the right-hand side of (2.3) by $sI_n - A$. The result is

$$(s^n + a_{n-1}s^{n-1} + \cdots + a_1 s)I_n - (a_1 A + a_2 A^2 + \cdots + A^n) \tag{2.4}$$

which by the Cayley–Hamilton theorem is $\Delta(s)I_n$. This proves (2.3).

Theorem 2.3 The inverse of $sI_n - A$ can be expanded as a matrix Laurent series (see below, Section 8)

$$(sI_n - A)^{-1} = s^{-1}I_n + s^{-2}A + s^{-3}A^2 + \cdots \tag{2.5}$$

convergent for all $|s| > \rho$, where ρ is $\max_i |\lambda_i|$ and the λ_i are the eigenvalues of A.

Proof This follows at once on writing

$$(sI_n - A)^{-1} = s^{-1}(I_n - s^{-1}A)^{-1} \tag{2.6}$$

and using the series expansion of $I_n - s^{-1}A$ (R & S, p. 186).

Theorem 2.4 (*Algorithm of Souriau–Frame–Faddeev*) If (2.3) is written

$$(sI_n - A)^{-1} = \frac{s^{n-1}I_n + s^{n-2}B_{n-2} + \cdots + sB_1 + B_0}{\Delta(s)} \tag{2.7}$$

then the matrices $B_0, B_1, \ldots, B_{n-2}$ and the coefficients $a_0, a_1, \ldots, a_{n-1}$ in $\Delta(s)$ can be computed by the following algorithm:

$$\left.\begin{array}{llll}
 & & B_{n-1} & \triangleq I_n \\
a_{n-1} & = -\text{trace}\,(AB_{n-1}), & B_{n-2} & = a_{n-1}I_n + AB_{n-1} \\
a_{n-2} & = -\tfrac{1}{2}\,\text{trace}\,(AB_{n-2}), & B_{n-3} & = a_{n-2}I_n + AB_{n-2} \\
\vdots & & \vdots & \\
a_1 & = -\dfrac{1}{n-1}\,\text{trace}\,(AB_1), & B_0 & = a_1I_n + AB_1 \\
a_0 & = -\dfrac{1}{n}\,\text{trace}\,(AB_0), & [0 & = a_0I_n + AB_0]
\end{array}\right\} \tag{2.8}$$

where the last equation serves as a check.

Proof Comparison of (2.3) and (2.7) shows that

$$\left.\begin{array}{l}
B_{n-2} = a_{n-1}I_n + A \\
B_{n-3} = a_{n-2}I_n + a_{n-1}A + A^2 = a_{n-2}I_n + AB_{n-2} \\
\vdots \\
B_1 = a_2I_n + a_3A + \cdots + A^{n-2} = a_2I_n + AB_2 \\
B_0 = a_1I_n + a_2A + \cdots + A^{n-1} = a_1I_n + AB_1
\end{array}\right\} \tag{2.9}$$

and by the Cayley–Hamilton theorem

$$a_0I_n + AB_0 = 0 \tag{2.10}$$

It remains to prove the formula for the a_i.

From (2.1) and (1.4) it readily follows that

$$a_{n-1} = -\text{trace } A \tag{2.11}$$

Now if H is non-singular,

$$|sI_n - H^{-1}AH| = |H^{-1}(sI_n - A)H| \tag{2.12}$$
$$= |sI_n - A| \tag{2.13}$$

Choose H so that $H^{-1}AH$ is in Jordan form J. Then equation (2.13) gives

$$|sI_n - J| = |sI_n - A| = \Delta \tag{2.14}$$

whence by (2.11)

$$\text{trace } A = \text{trace } J = \sum_{i=1}^{n} \lambda_i \triangleq \sigma_1 \tag{2.15}$$

where the λ_i are the eigenvalues of A. In the same way

$$\text{trace } A^r = \text{trace } J^r = \sum_{i=1}^{n} \lambda_i^r \triangleq \sigma_r, \quad r = 1, 2, \ldots \tag{2.16}$$

Now from (2.14), as in (1.57),

$$\prod_{i=1}^{n} (s - \lambda_i) = s^n + a_{n-1}s^{n-1} + \cdots + a_1 s + a_0 \tag{2.17}$$

and using Newton's formulae (see below, Theorem 3.2) we have by (2.16) and (2.9)

$$ka_{n-k} = -(\sigma_k + a_{n-1}\sigma_{k-1} + \cdots + a_{n-k+1}\sigma_1) \tag{2.18}$$
$$= -\text{trace } (A^k + a_{n-1}A^{k-1} + \cdots + a_{n-k+1}A) \tag{2.19}$$
$$= -\text{trace } (AB_{n-k}), \quad k = 1, 2, \ldots, n \tag{2.20}$$

which completes the proof.

The Cayley–Hamilton theorem shows that $\Delta(A) = 0$, but there may be polynomials $p(s)$ of lower degree than $\Delta(s)$ such that $p(A) = 0$. The monic polynomial of lowest degree, $m(s)$, such that $m(A) = 0$ is called the *minimal polynomial* of A. It is a divisor of $\Delta(s)$, and so we may write

$$\Delta(s) = q(s)m(s) \tag{2.21}$$

where $q(s)$ is a monic polynomial. Then it can be shown that the polynomial matrix adj $(sI_n - A)$, which is equal to $\Delta(s)(sI_n - A)^{-1}$, has $q(s)$ as the monic greatest common divisor of its elements (R & S, p. 89). Therefore

$$\text{adj } (sI_n - A) = q(s)\Gamma(s) \tag{2.22}$$

where the *reduced adjoint matrix* $\Gamma(s)$ has elements with no polynomial common factor. Also (2.21) and (2.22) give

$$(sI_n - A)^{-1} = \frac{\text{adj}\,(sI_n - A)}{\Delta(s)} = \frac{\Gamma(s)}{m(s)} \tag{2.23}$$

In later chapters we shall consider the *transfer function matrix* $G(s)$,

$$G(s) = C(sI_n - A)^{-1}B = \frac{C\Gamma(s)B}{m(s)} \tag{2.24}$$

where C is $m \times n$, B is $n \times l$. Though $m(s)$ has no factor in common with every element of $\Gamma(s)$, it may have such a factor in common with every element of $C\Gamma(s)B$ (R & S, p. 90).

3 Polynomials

Polynomials have already been mentioned above. They can be regarded from two points of view. A *polynomial form* is an expression

$$p(s) = a_0 + a_1 s + \cdots + a_n s^n \tag{3.1}$$

in which $a_0, a_1, \ldots, a_n$ belong to a field F (more generally, a ring R) and s is an *indeterminate* such that s^r and as are defined whenever $a \in F$; and $as = sa$. If $a_n \neq 0$, the number n is the *degree* of $p(s)$, denoted by $\delta(p)$. If $a_n = 1$, the polynomial is said to be *monic*. The polynomial form

$$q(s) = b_0 + b_1 s + \cdots + b_n s^n \tag{3.2}$$

is equal to $p(s)$ if and only if

$$a_i = b_i, \quad i = 0, 1, \ldots, n \tag{3.3}$$

Notice that no question arises in polynomial forms of giving a value in F to s.

A *polynomial function* is an expression

$$p(s) = a_0 + a_1 s + \cdots + a_n s^n \tag{3.4}$$

in which now $a_0, a_1, \ldots, a_n$ and s all belong to F (more generally, R). Therefore $p(s)$ associates, with each member s_0 of F, another member $p(s_0)$. When the field F is infinite, no distinction need be drawn between polynomial forms and polynomial functions. In this book polynomial functions are used throughout, because the important fields F are the real or the complex numbers, both infinite. To deal with finite fields, if this were ever wanted, it would be necessary in a few places to substitute polynomial forms for polynomial functions and make appropriate changes in the proofs of theorems.

The values of s for which a polynomial function $p(s)$ takes the value $0 \in F$ are called the *roots* or *zeros* of p. A polynomial $g(s)$ *divides* a polynomial $p(s)$ if there exists a polynomial $p_1(s)$ such that

$$p(s) = g(s)p_1(s) \tag{3.5}$$

If $g(s)$ divides $p(s)$ and $q(s)$, but no polynomial of higher degree than $g(s)$ divides $p(s)$ and $q(s)$, then $g(s)$ is a *greatest common divisor* of $p(s)$ and $q(s)$. If a greatest common divisor of $p(s)$ and $q(s)$ has degree zero, then $p(s)$ and $q(s)$ are *relatively prime.*

Theorem 3.1 The two polynomials with real or complex coefficients

$$p(s) = p_0 + p_1 s + \cdots + p_n s^n \tag{3.6}$$

$$q(s) = q_0 + q_1 s + \cdots + q_m s^m \tag{3.7}$$

are relatively prime if and only if the following $(m + n) \times (m + n)$ matrix R has rank $m + n$ (that is, $|R| \neq 0$),

$$R = \begin{pmatrix} p_n & 0 & \cdots & 0 & 0 & 0 & 0 & \cdots & 0 & q_m \\ p_{n-1} & p_n & \cdots & 0 & 0 & 0 & 0 & \cdots & q_m & q_{m-1} \\ p_{n-2} & p_{n-1} & \cdots & 0 & 0 & 0 & 0 & \cdots & q_{m-1} & q_{m-2} \\ \vdots & \vdots & & \vdots & \vdots & \vdots & \vdots & & \vdots & \vdots \\ 0 & 0 & \cdots & p_0 & p_1 & q_1 & q_0 & \cdots & 0 & 0 \\ 0 & 0 & \cdots & 0 & p_0 & q_0 & 0 & \cdots & 0 & 0 \end{pmatrix} \tag{3.8}$$

The determinant of R is called the *resultant* of $p(s)$ and $q(s)$, and R will be called the *resultant matrix.*

Proof Suppose that there is a common root α,

$$p(s) = (s - \alpha)(-a_0 - a_1 s - \cdots - a_{n-1}s^{n-1}) \tag{3.9}$$

$$q(s) = (s - \alpha)(b_0 + b_1 s + \cdots + b_{m-1}s^{m-1}) \tag{3.10}$$

Then eliminating $s - \alpha$,

$$p(s)(b_0 + b_1 s + \cdots + b_{m-1}s^{m-1}) + q(s)(a_0 + a_1 s + \cdots + a_{n-1}s^{n-1}) = 0 \tag{3.11}$$

and using (3.6), (3.7) we have, on equating the coefficients of each power of s to zero,

$$\begin{aligned} p_n b_{m-1} \qquad\qquad\qquad\qquad + q_m a_{n-1} &= 0 \\ p_{n-1}b_{m-1} + p_n b_{m-2} + q_m q_{n-2} + q_{m-1}q_{n-1} &= 0 \\ \vdots \qquad\qquad\qquad &\ \ \vdots \\ p_0 b_0 + q_0 a_0 \qquad &= 0 \end{aligned} \tag{3.12}$$

It follows from Section 1.2 that these equations have a non-zero solution $a_0, a_1, \ldots, a_{n-1}, b_0, b_1, \ldots, b_{m-1}$ if and only if the determinant of the coefficient matrix is zero: this matrix is R.

Theorem 3.2 (*Newton's formulae*) Let the roots of the polynomial

$$s^n + a_{n-1}s^{n-1} + \cdots + a_1 s + a_0 \tag{3.13}$$

be $\lambda_1, \lambda_2, \ldots, \lambda_n$. Write

$$\sigma_r = \sum_{i=1}^{n} \lambda_i^r, \quad r = 1, 2, \ldots, n \tag{3.14}$$

Then

$$\sigma_k + a_{n-1}\sigma_{k-1} + \cdots + a_{n-k+1}\sigma_1 + ka_{n-k} = 0, \quad k = 1, 2, \ldots, n \tag{3.15}$$

Proof We have

$$f(s) = s^n + a_{n-1}s^{n-1} + \cdots + a_1 s + a_0 \tag{3.16}$$

$$= \prod_{i=1}^{n} (s - \lambda_i) \tag{3.17}$$

whence two expressions for the derivative can be obtained,

$$f'(s) = ns^{n-1} + (n-1)a_{n-1}s^{n-2} + \cdots + a_1 \tag{3.18}$$

$$= \sum_{i=1}^{n} \frac{f(s)}{s - \lambda_i} \tag{3.19}$$

Now for any λ, the division algorithm gives

$$\frac{f(s)}{s-\lambda} = s^{n-1} + (\lambda + a_{n-1})s^{n-2} + (\lambda^2 + a_{n-1}\lambda + a_{n-2})s^{n-3}$$
$$+ \cdots + (\lambda^{n-1} + a_{n-1}\lambda^{n-2} + \cdots + a_2\lambda + a_1) + \frac{f(\lambda)}{s-\lambda} \tag{3.20}$$

and adding n such equations for $\lambda = \lambda_1, \lambda_2, \ldots, \lambda_n$ we obtain, because $f(\lambda_i) = 0$,

$$f'(s) = ns^{n-1} + (\sigma_1 + na_{n-1})s^{n-2} + (\sigma_2 + a_{n-1}\sigma_1 + na_{n-2})s^{n-3}$$
$$+ \cdots + (\sigma_{n-1} + a_{n-1}\sigma_{n-2} + \cdots + a_2\sigma_1 + na_1) \tag{3.21}$$

Then equating (3.18) and (3.21) we obtain (3.15) for $k = 1, 2, \ldots, n-1$. But

$$\lambda_i^n + a_{n-1}\lambda_i^{n-1} + \cdots + a_1\lambda_i + a_0 = 0, \quad i = 1, 2, \ldots, n \tag{3.22}$$

and on adding these n equations we obtain (3.15) for $k = n$.

Just as the set of real integers can be embedded in the set of rational numbers, so the polynomials can be embedded in the *rational polynomials*. From the polynomial forms we thus generate *rational forms*, and from the polynomial functions we generate *rational functions*. Again we need make no distinction between these when the coefficients are taken from an infinite field.

4 Linear vector spaces

A linear vector space over a field F is a set of *vectors* v_i which are undefined except that they must have the following properties. Multiplication of v_i by a scalar $\alpha \in F$ must be defined, and is written αv_i, and $\alpha v_i \in V$. Addition of v_i and v_j must be defined, and is written $v_i + v_j$, and $v_i + v_j \in V$. The vectors v_i form a group under addition, and

$$\alpha(v_i + v_j) = \alpha v_i + \alpha v_j \tag{4.1}$$

$$(\alpha + \beta)v_i = \alpha v_i + \beta v_i \tag{4.2}$$

$$\alpha(\beta v_i) = (\alpha\beta)v_i \tag{4.3}$$

$$1v_i = v_i \tag{4.4}$$

All these properties hold for all $\alpha, \beta \in F$ and all $v_i, v_j \in V$. Examples are physical space, and sets of n-tuples called *column vectors*

$$x = \begin{pmatrix} x_1 \\ x_2 \\ \vdots \\ x_n \end{pmatrix} \tag{4.5}$$

with the *elements* $x_1, x_2, \ldots, x_n$ of the vector x being members of F; and with addition and multiplication defined by

$$x + y = \begin{pmatrix} x_1 + y_1 \\ x_2 + y_2 \\ \vdots \\ x_n + y_n \end{pmatrix}, \qquad \alpha x = \begin{pmatrix} \alpha x_1 \\ \alpha x_2 \\ \vdots \\ \alpha x_n \end{pmatrix} \tag{4.6}$$

A linear vector space is *finite-dimensional* if every vector $x \in V$ can be written

$$x = \alpha_1 v_1 + \alpha_2 v_2 + \cdots + \alpha_n v_n \tag{4.7}$$

where $v_1, v_2, \ldots, v_n$ are fixed vectors in V. If vectors $w_1, w_2, \ldots, w_m$ are such that no $\beta_i \in F$ exist (other than zero) which make

$$\beta_1 w_1 + \beta_2 w_2 + \cdots + \beta_m w_m = 0 \tag{4.8}$$

then $w_1, w_2, \ldots, w_m$ are *linearly independent*; otherwise they are *linearly dependent*. If the vectors v_i in (4.7) are linearly independent they form a *basis* for V, and V is n-dimensional.

A *linear transformation* A from a linear vector space V to another linear vector space W establishes a correspondence between any $v \in V$ and a unique $w = A(v) \in W$, and has the properties

$$A(v_i + v_j) = A(v_i) + A(v_j) \tag{4.9}$$

$$A(\alpha v_i) = \alpha A(v_i) \tag{4.10}$$

The *image* $A(v)$ of v under A is often written Av. Given a basis for V, any linear transformation A can be represented by a matrix. Usually the matrix is also given the symbol A.

Given any r vectors $v_1, v_2, \ldots, v_r \in V$, we may generate all possible vectors w

$$w = \alpha_1 v_2 + \alpha_2 v_2 + \cdots + \alpha_r v_r \tag{4.11}$$

where $\alpha_1, \alpha_2, \ldots, \alpha_r \in F$. The set of all such w (for different α_i) constitute a *subspace* W of V: as a special case W may be equal to V. The subspace W is said to be *spanned* by the vectors $v_1, v_2, \ldots, v_r$.

Given any $m \times n$ matrix A over a field, its columns can be regarded as vectors in an m-dimensional space, and its rows as vectors in an n-dimensional space. The number of linearly independent columns of A is called its *column rank*, and the number of linearly independent rows is called its *row rank*. With determinantal rank defined as in Section 1 we have, for any matrix.

$$\text{determinantal rank} = \text{row rank} = \text{column rank} \tag{4.12}$$

We may therefore talk simply of *rank* to mean all of these (R & S, p. 78).

According to the usual rule of matrix multiplication, if $v \in V$ is a column vector then Av is also a column vector, which is a sum of multiples (by the elements of v) of the columns of A. The subspace spanned by the columns of A is called its *range* or *column space*, and its dimension is the rank of A. The set of vectors v satisfying.

$$Av = 0 \tag{4.13}$$

also define a subspace which is called the *kernel* or *null space* of A, and its dimension is the *nullity* of A. We have (R & S, pp. 71–72),

$$\text{nullity } A + \text{rank } A = \dim V \tag{4.14}$$

whence if A is $m \times n$ with $n \leqslant m$, the nullity is the same as the rank defect (Section 1), but otherwise not.

If v is an eigenvector of A we have by equation (1.59)

$$Av = \lambda v \tag{4.15}$$

Therefore Av lies in the subspace spanned by v: this subspace is said to be *invariant under A*. More generally, if $v_1, v_2, \ldots, v_r$ span a subspace W of V, and if $Av_1, Av_2, \ldots, Av_r$ lie in the same subspace W, then W is invariant under A.

In later chapters we shall be concerned with vectors over the following fields:

(i) The field of real numbers.
(ii) The field of complex numbers.
(iii) The field of rational functions.

We shall also meet cases where the elements $v_i(s)$ of a column vector $v(s)$ are polynomials in s and we multiply these by other polynomials. We then regard $v(s)$ as being embedded in a vector space over the field of rational functions (that is, the scalars α, β in (4.1) to (4.3) are rational functions). In particular we have the following theorem.

Theorem 4.1 The polynomial column vectors $w_1(s), w_2(s), \ldots, w_n(s)$ are linearly dependent if and only if there exist polynomials $\alpha_1(s), \alpha_2(s), \ldots, \alpha_n(s)$ such that

$$\alpha_1(s)w_1(s) + \alpha_2(s)w_2(s) + \cdots + \alpha_n(s)w_n(s) \tag{4.16}$$

is the zero polynomial.

Proof The vectors $w_i(s)$ belong to a vector space over the field of rational functions. They are linearly dependent if and only if there exist rational functions $\beta_i(s)$ such that

$$\beta_1(s)w_1(s) + \beta_2(s)w_2(s) + \cdots + \beta_n(s)w_n(s) \tag{4.17}$$

is the zero rational function. Let the product of the denominator polynomials of $\beta_1(s), \beta_2(s), \ldots, \beta_n(s)$ be $p(s)$, so that the $p(s)\beta_i(s)$ are polynomials. Then the sum (4.16) is the zero rational function if and only if

$$p(s)\beta_1(s)w_1(s) + p(s)\beta_2(s)w_2(s) + \cdots + p(s)\beta_n(s)w_n(s) \tag{4.18}$$

is the zero polynomial. On putting $p(s)\beta_i(s) = \alpha_i(s)$ this gives the desired result.

In Chapter 6 we shall also meet column vectors $v(t)$ having as their elements real or complex functions of t. These, however, offer no difficulty as the scalars are the real or complex numbers.

Let the vectors of a linear vector space V over the real numbers be n-tuples of real numbers as in (4.5). If we define an *inner product* on this space by

$$x^{\mathrm{T}}y = x_1y_1 + x_2y_2 + \cdots + x_ny_n \tag{4.19}$$

then V becomes *euclidean n-space*, denoted by E_n. The *euclidean norm* of a vector $x \in E_n$ is

$$\|x\| = \sqrt{\{x^{\mathrm{T}}x\}} \geqslant 0 \tag{4.20}$$

A *neighbourhood* $N(x, h)$ of a point xE_n is the set

$$N(x, h) = \{y \mid \|y - x\| < h\} \tag{4.21}$$

When the value of h is of no consequence, $N(x, h)$ is written $N(x)$. A *deleted neighbourhood* $N'(x, h)$ of x is the set

$$N'(x, h) = \{y \mid \|y - x\| < h \text{ and } y \neq x\} \tag{4.22}$$

5 Limiting processes

The ideas of a limit, of continuity, of the derivative, etc. will be assumed to be familiar (R & S, Chapter 4). Some definitions and a few of the less usual or non-standard results are collected here.

An *open set* S in E_n is a set such that every point in S has a neighbourhood also in S. A *closed set* S in E_n is such that if any sequence of vectors $x_1, x_2, \ldots$ in S tends to a limit x, then $x \in S$.

A function f having as domain the real numbers and as range E_m is *continuous on the left at t* if

$$\lim_{\substack{t \to t_0 \\ t < t_0}} f(t) = f(t_0) \tag{5.1}$$

Continuity on the right is defined in a similar way.

The function f has a *left-hand derivative* $D_{-}f(t_0)$ at t_0 if

$$\lim_{\substack{t \to t_0 \\ t < t_0}} \frac{f(t) - f(t_0)}{t - t_0} \tag{5.2}$$

exists and is finite, in which case the limit is equal to $D_{-}f(t_0)$. The *right-hand derivative* $D_{+}f(t_0)$ is defined in a similar way. If $D_{-}f(t_0)$ and $D_{+}f(t_0)$ exist and are equal, then f has a *derivative* $Df(t_0)$ at t_0 equal to this common value.

If f is a function with domain $D \in E_n$ and range $R \in E_m$, the derivative Df of f, if it exists, is an $m \times n$ matrix

$$Df(x_0) = \begin{pmatrix} \dfrac{\partial f_1}{\partial x_1} & \dfrac{\partial f_1}{\partial x_2} & \cdots & \dfrac{\partial f_1}{\partial x_n} \\ \dfrac{\partial f_2}{\partial x_1} & \dfrac{\partial f_2}{\partial x_2} & \cdots & \dfrac{\partial f_2}{\partial x_n} \\ \vdots & \vdots & & \vdots \\ \dfrac{\partial f_m}{\partial x_1} & \dfrac{\partial f_m}{\partial x_2} & \cdots & \dfrac{\partial f_m}{\partial x_n} \end{pmatrix} \tag{5.3}$$

where $\partial f_i/\partial x_j$ is the partial derivative of the element f_i of f with respect to the element x_j of x, evaluated at x_0. The existence of the partial derivatives at x_0 does not imply the existence of $Df(x_0)$ (R & S, pp. 148, 149), but the existence and continuity of the partial derivatives in a neighbourhood of x_0 does imply the existence of $Df(x_0)$ (R & S, p. 157). Another notation for the partial derivative $\partial f_i/\partial x_j$ of $f_i(x_1, x_2, \ldots, x_n)$ is $D_j f_i$.

Let h be the compound function gf, so that $h(x) = g[f(x)]$. Then we have the following rule for obtaining the derivative of h.

Theorem 5.1 (The *chain rule*) Let f be a continuous function with domain an open neighbourhood $N(x_0) \subseteq E_n$ and range in E_m. Let $y_0 = f(x_0)$ and let g be a continuous function with domain an open neighbourhood $N(y_0) \subseteq E_m$ and range in E_l. If f is differentiable at x_0 and g is differentiable at y_0 then the composite function $h = gf$ is differentiable at x_0 and $Dh = Dg\, Df$, which is to be regarded as the product of two matrices.

Proof See R & S, pp. 150, 151.

All physical variables in this book will be assumed to be continuous on the left (R & S, pp. 142–144). In fact they will be assumed to belong to the class D_R or the class D_R^1 as appropriate (R & S, pp. 167–168). These are defined in the following way.

(i) The function f belongs to D_R on an open interval I if it can be written on any bounded open interval $J \subset I$ in the form

$$f(t) = \sum_{i=1}^{p} f_i U(t - p_i) + \phi(t) \tag{5.4}$$

where the $p_i \in J$ and the f_i are constants, and ϕ is continuous and bounded on J. Here $U(t - p_i)$ represents the *unit step* at p_i, defined by

$$U(t - p_i) = \begin{cases} 0, & t \leqslant p_i \\ 1, & t > p_i \end{cases} \tag{5.5}$$

which is therefore continuous on the left but discontinuous on the right at p_i. The class D_L is defined in a similar way, but with $U(t - p_i)$ replaced by $U(p_i - t)$.

(ii) The function f belongs to D_R^1 on an open interval I if on any bounded open interval $J \subset I$ it can be written

$$f(t) = \sum_{i=1}^{p} f_i U(t - p_i) + \sum_{i=1}^{q} (t - q_i)\phi_i U(t - q_i) + \psi(t) \tag{5.6}$$

where the $p_i \in J$, $q_i \in J$, f_i, and ϕ_i are constants, and ψ is continuous and bounded on J and has continuous and bounded derivative $D\psi$ on J. The class D_L^1 is defined similarly, with $U(t - p_i)$, $U(t - q_i)$ replaced by $U(p_i - t)$, $U(q_i - t)$.

If $f \in D_R^1$, then clearly $f \in D_R$, and similarly if $f \in D_L^1$ then $f \in D_L$. The definitions apply to vector-valued functions f as well as real-valued functions. When I is not specified it is understood to be $(-\infty, \infty)$.

6 Integration

The theory of the Riemann–Stieltjes integral will be assumed (R & S, Chapter 5). Only a few of the less familiar results will be given.

Theorem 6.1 Let f and α be real-valued functions defined on an open interval I, and let $[a, b]$ be a closed interval contained in I. If $f \in D_R^1$ on I and $\alpha \in D_L^1$ on I, then the Riemann–Stieltjes integral

$$\int_a^b f(t)\, d\alpha(t) \tag{6.1}$$

exists. The integral also exists if $f \in D_L^1$, $\alpha \in D_R^1$ on I.

Proof See R & S, pp. 202–203.

If f is vector-valued, then the integral (6.1) is defined to mean

$$\int_a^b f\, d\alpha = \begin{pmatrix} \int_a^b f_1\, d\alpha \\ \int_a^b f_2\, d\alpha \\ \vdots \\ \int_a^b f_n\, d\alpha \end{pmatrix} \tag{6.2}$$

If f and α are both vector-valued (with n elements) then we define

$$\int_a^b f^{\mathrm{T}}\, d\alpha = \sum_{i=1}^{n} \int_a^b f_i\, d\alpha_i \tag{6.3}$$

provided that each of the integrals in the sum exists. Similarly if A is an $m \times n$ matrix-valued function, and B is $n \times l$, we define

$$\int_a^b A\, dB \tag{6.4}$$

as the matrix having as ijth element

$$\sum_{k=1}^{n} \int_a^b a_{ik}\, db_{kj} \tag{6.5}$$

provided that each integral in the sum exists.

Theorem 6.2 (*Integration by parts*) If the integral $\int_a^b A\,dB$ exists, then $\int_a^b B^{\mathrm{T}}\,dA^{\mathrm{T}}$ exists and is given by

$$\left\{\int_a^b B^{\mathrm{T}}\,dA^{\mathrm{T}}\right\}^{\mathrm{T}} = A(b)B(b) - A(a)B(a) - \int_a^b A\,dB \tag{6.6}$$

We also write (when no confusion is likely)

$$\left\{\int_a^b B^{\mathrm{T}}\,dA^{\mathrm{T}}\right\}^{\mathrm{T}} = \int dA.B \tag{6.7}$$

Proof See R & S, pp. 195–198.

When the integrator $\alpha(t)$ becomes simply t, the Riemann–Stieltjes integral becomes the Riemann integral

$$\int_a^b f(t)\,dt \tag{6.8}$$

This integral exists (R & S, p. 202) if either

(i) f is continuous on $[a, b]$; or
(ii) f has bounded variation on $[a, b]$.

When the integral (6.8) exists, the value of $f(t)$ may be changed (to any other finite value) at any finite set of points in $[a, b]$ without changing the value of the integral (R & S, p. 203).

Theorem 6.3 Let f be a vector-valued function, continuous on $[a, b]$. Then the function F defined by

$$F(x) = \int_a^x f(t)\,dt \tag{6.9}$$

is continuous on $[a, b]$ and has continuous first derivative $DF(x) = f(x)$ on (a, b).

Proof See R & S, pp. 204, 223–224.

7 Differential equations

Let x be an n-vector and t a scalar, and let $f(x, t)$ be defined on the region R,

$$R = \{x, t \mid \|x - c\| \leqslant \alpha, \beta_1 \leqslant t - t_0 \leqslant \beta_2\} \tag{7.1}$$

where $0 \in [\beta_1, \beta_2]$. Then we say $f \in (C, \mathrm{Lip})$ if

(i) f is a continuous function of t on R; and

(ii) f satisfies a *Lipschitz condition* with respect to x on R: that is, there exists a $k > 0$ such that for every $(x, t) \in R$ and $(y, t) \in R$

$$\|f(y, t) - f(x, t)\| \leqslant k\|y - x\| \tag{7.2}$$

Now suppose that we are given a differential equation

$$\dot{x}(t) = f[x(t), t] \tag{7.3}$$

and an initial condition $x(t_0) = c$. Then we have the following existence theorems.

Theorem 7.1 If $f \in (C, \text{Lip})$ on R, there is a unique solution of (7.3) passing through (c, t_0) and remaining in R on a non-zero interval $[\gamma_1, \gamma_2]$ containing t_0. This solution is written $\phi(t; c, t_0)$; it is continuous on $[\gamma_1, \gamma_2]$ and has continuous first derivative on (γ_1, γ_2).

Proof See R & S, pp. 245–246.

Theorem 7.2 Let A, B, u be continuous functions of t on $\beta_1 \leqslant t - t_0 \leqslant \beta_2$, where $0 \in [\beta_1, \beta_2]$. Then there exists a solution $\phi(t; c, t_0)$ of the differential equation

$$\dot{x}(t) = A(t)x(t) + B(t)u(t) \tag{7.4}$$

passing through (c, t_0). This solution ϕ is unique, is continuous on $\beta_1 \leqslant t - t_0 \leqslant \beta_2$, and has continuous first derivative on $\beta_1 < t - t_0 < \beta_2$.

Proof See R & S, pp. 251–252.

Theorem 7.3 Under the conditions of Theorem 7.2, let Ψ be the unique solution of the matrix differential equation

$$\dot{\Psi}(t) = A(t)\Psi(t) \tag{7.5}$$

which takes the value I_n at $t = t_0$. Then the solution ϕ defined in Theorem 7.2 is given by

$$\phi(t; c, t_0) = \Psi(t)c + \Psi(t)\int_{t_0}^{t} \Psi^{-1}(\tau)B(\tau)u(\tau)\, d\tau \tag{7.6}$$

Proof See R & S, p. 253.

The matrix Ψ is called the *transition matrix*, and is usually written $\Psi(t, t_0)$ to emphasize its dependence on t_0. It is non-singular for all t, t_0 having

$$\beta_1 \leqslant t - t_0 \leqslant \beta_2.$$

Theorem 7.4 Let A be a constant $n \times n$ matrix and B a constant $n \times l$ matrix. Let $u \in D_R$ be a vector-valued function of t with l elements, and let c be a constant n-vector. Then there exists a unique continuous solution $\phi(t; c, t_0)$ of

$$D_-\phi(t) = A\phi(t) + Bu(t) \tag{7.7}$$

passing through (c, t_0). If u is continuous at $t = t_1$, then $D\phi(t_1)$ exists. The function ϕ is given explicitly by

$$\phi(t; c, t_0) = e^{A(t-t_0)}c + \int_{t_0}^{t} e^{A(t-\tau)}Bu(\tau)\,d\tau \tag{7.8}$$

Proof See R & S, pp. 237–240.

8 Complex functions

If f is a complex-valued function of the complex variable z, then f has a derivative $f'(z_0)$ at z_0 if the limit

$$\lim_{z\to z_0} \frac{f(z) - f(z_0)}{z - z_0} \tag{8.1}$$

exists and is finite, and the limit is then equal to $f'(z_0)$.

Write $z = x + iy, f = u + iv$, and define real 2-vectors

$$w = \begin{pmatrix} x \\ y \end{pmatrix}, \qquad g = \begin{pmatrix} u \\ v \end{pmatrix} \tag{8.2}$$

The variables $\operatorname{Re} z = x, \operatorname{Im} z = y$ are treated as coordinates in the *complex plane*, and the *modulus* $|z|$ is defined to be the euclidean norm $\|w\|$ of w. The *argument* of z, $\arg z$, is an angle θ (in radians) such that $x = |z| \cos \theta$, $y = |z| \sin \theta$, and the *principal value* of $\arg z$ is that value of θ satisfying $-\pi < \theta \leqslant \pi$.

Existence of the limit (8.1) implies the existence of the derivative $Dg(w_0)$ of g regarded as a function of the real 2-vector w (Dg is a 2×2 matrix). It implies also that the partial derivatives occurring in Dg satisfy the *Cauchy–Riemann equations* at w_0,

$$\begin{aligned} \frac{\partial u}{\partial x} &= \frac{\partial v}{\partial y} \\ \frac{\partial v}{\partial x} &= -\frac{\partial u}{\partial y} \end{aligned} \tag{8.3}$$

If f is defined on an open set S in the complex plane, and if f' exists everywhere in S, then f is said to be *analytic* on S. If f is analytic on $N(z_0)$, it is said to be *analytic at* z_0.

Theorem 8.1 If f is analytic inside and on the boundary of the annulus

$$r \leqslant |z - z_0| \leqslant R \tag{8.4}$$

then $f(z)$ is given by the *Laurent series*

$$f(z) = \sum_{n=0}^{\infty} a_n(z - z_0)^n + \sum_{n=1}^{\infty} b_n(z - z_0)^{-n} \tag{8.5}$$

Proof See R & S, pp. 279–280, where the formulae for a_n and b_n are also given.

If f is analytic at z_0, every b_i is zero and the Laurent series is called a *Taylor series*. Then if $a_0, a_1, \ldots, a_{m-1}$ are zero, but $a_m \neq 0$ for some $m > 0$, f is said to have a zero of order m at z_0. Suppose that f is not analytic at z_0, but is analytic in a deleted neighbourhood $N'(z_0)$. Construct an annular region centred on z_0 and contained in $N'(z_0)$. If in the Laurent series $b_m \neq 0$, but $b_{m+1}, b_{m+2}, \ldots$ are all zero, then f is said to have a *pole of order m* at z_0.

An *elementary contour C* is a curve made up of straight-line segments and of circular arcs. A *closed elementary contour* does not cross itself but divides the complex planes into two regions, interior and exterior to the contour. A *contour integral* $\int_C f(z)\,dz$ is the complex Riemann–Stieltjes integral obtained by taking z on C as a suitable function of a real parameter t and defining

$$\int_C f(z)\,dz = \int_{t_a}^{t_b} f[z(t)]dz(t) \tag{8.6}$$

where t_a, t_b correspond to the beginning and the end of the contour C.

Theorem 8.2 If f is analytic inside and on a closed elementary contour C, except at a finite number of poles inside C, and if $f(z)$ does not vanish on C, then

$$\frac{1}{2\pi i}\int_C \frac{f'(z)}{f(z)}\,dz = Z - P \tag{8.7}$$

where Z is the number of zeros inside C, and P is the number of poles (each pole or zero of order m being counted m times).

Proof See R & S, pp. 286–287.

Corollary (*Principle of the argument*) As z goes once clockwise (resp. anticlockwise) around C, let $f(z)$ trace out the closed curve Γ. Then Γ encircles the origin $Z - P$ times in the clockwise (resp. anticlockwise) direction.

Proof From equation (8.7),

$$Z - P = \frac{1}{2\pi i}\int_C d\{\log f(z)\} \tag{8.8}$$

$$= \frac{1}{2\pi i}\int_C d\{\log |f(z)| + i \arg f(z)\} \tag{8.9}$$

When z goes once anticlockwise round C, $|f(z)|$ returns to its original value. The value of $\arg f(z)$ increases by 2π times the number of encirclements by Γ of the origin.

If $\operatorname{Re} z \geqslant 0$ we shall say that z lies in the *closed right half-plane*. If $\operatorname{Re} z > 0$ we shall say that z is in the *open right half-plane*. The open and closed left half-planes are defined in the corresponding way.

9 The Laplace transform

The relation

$$\bar{f}(s) = \int_0^\infty e^{-st} f(t)\, dt \tag{9.1}$$

sets up a one-to-one relation between a suitable real-valued function f of the real variable t, and a complex-valued function $\bar{f}$ of the complex variable s (R & S, pp. 303–309). Suitable conditions on the function f in equation (9.1) are (R & S, p. 295):

(i) $f \in D^1_R$, and
(ii) $|f(t)| < M_1 e^{\alpha_1 t}$, $t > 0$, and
(iii) $f(t) = 0$, $t \leq 0$

If f is vector-valued or matrix-valued (9.1) is interpreted in the natural way. For example if f in (9.1) is an $m \times n$ matrix, then $\bar{f}$ is an $m \times n$ matrix having as its ijth element

$$\bar{f}_{ij}(s) = \int_0^\infty e^{-st} f_{ij}(t)\, dt \tag{9.2}$$

Theorem 9.1 If the Laplace transform of f_1 is $\bar{f}_1$, and of f_2 is $\bar{f}_2$, then the Laplace transform of $\alpha_1 f_1 + \alpha_2 f_2$ is $\alpha_1 \bar{f}_1 + \alpha_2 \bar{f}_2$. Here α_1, α_2 are real (resp. complex) constants.

Proof This follows from (9.1) and the properties of the integral.

Let r vector-valued functions f_i be given, and from them generate the linear vector space V,

$$V = \{f \mid f = \sum_{i=1}^{r} \beta_i f_i\} \tag{9.3}$$

where the β_i are any real (resp. complex) constants. Similarly, generate a linear vector space $\bar{V}$ from the $\bar{f}_i$. Then by Theorem 9.1 and the one-to-one nature of the Laplace transform, V and $\bar{V}$ are *isomorphic* (R & S, p. 61). In particular, V and $\bar{V}$ have the same dimension.

Theorem 9.2 Let A, B be constant matrices, respectively $n \times n$, $n \times l$. Let $u \in D_R^1$ be an l-vector function of t satisfying $u(t) = 0$, $t \leqslant 0$, and $\|u(t)\| < Me^{\alpha t}$ for $t > 0$. Then the solution of the differential equation

$$\dot{x}(t) = Ax(t) + Bu(t); \qquad x(0) = 0 \tag{9.4}$$

is

$$x(t) = \begin{cases} 0, & t \leqslant 0 \\ \int_0^t e^{(t-\tau)A} Bu(\tau)\, d\tau, & t > 0 \end{cases} \tag{9.5}$$

and has Laplace transform

$$\bar{x}(s) = (sI_n - A)^{-1} B\bar{u}(s) \tag{9.6}$$

Proof See R & S, pp. 312–314.

Note that x defined by (9.5) satisfies (9.4) in the sense that for all t it is continuous and satisfies

$$D_- x(t) = Ax(t) + Bu(t) \tag{9.7}$$

For each t where u is continuous, $\dot{x}(t)$ exists and satisfies (9.4), but where u is discontinuous $\dot{x}(t)$ does not exist.

The *z-transform* (R & S, p. 322) can be regarded as a special case of the Laplace transform. Given a sequence $x_0, x_1, x_2, \ldots$, the z-transform is defined by

$$\tilde{x}(z) = \sum_{r=0}^{\infty} x_r z^{-r} \tag{9.8}$$

where it exists for complex z.

10 Dynamical systems

A dynamical system is more easily recognized than defined. For our purposes it is a physical system (the *plant*) to which is applied an *input* u and which responds by producing an *output* y. The input u and output y are real vector-valued functions of time, respectively an l-vector and an m-vector. A multivariable system is one in which $l > 1$ or $m > 1$. The way in which y is generated within the system from u can in general be very complicated, but we are concerned with two important cases.

(i) A *differential system* is one governed by a set of ordinary differential equations, typically

$$\dot{x}(t) = f[x(t), t, u(t)] \tag{10.1}$$

$$y(t) = g[x(t), t, u(t)] \tag{10.2}$$

In order to solve (10.1) it is necessary to give initial conditions, and often these are $x(t_0) = 0$. It is then usual to specify that $u(t) = 0$, $t \leqslant t_0$. With suitable conditions on the functions f, u, equation (10.1) then has a unique solution for $t > 0$.

(ii) A *discrete-time system* is governed by a difference equation, typically

$$x_{k+1} = f_k(x_k, u_k) \tag{10.3}$$

$$y_k = g_k(x_k, u_k) \tag{10.4}$$

Again we often have $x_0 = 0$ and $u_k = 0$, $k < 0$.

In (10.1) and (10.3), x is an n-vector termed the *state*. Clearly the dimension n of x is equal to the number of independent initial conditions which must be imposed to obtain a unique solution of (10.1) or (10.3). For example, when we write $x(t_0) = 0$, this specifies the n elements of $x(t_0)$, and we could give any other values to these elements. The number n is called the order of the system.

When the defining equations are not first-order differential or difference equations, as in (10.1), (10.3), the definition of the system order is less straightforward. The question will be examined in detail in Chapter 5, Section 8. For the present, however, we adopt the following as a working definition: the *order* of a system is the number of independent initial conditions which must be specified to obtain a unique solution.

A *linear time-dependent differential system* obeys linear equations such as

$$\dot{x}(t) = A(t)x(t) + B(t)u(t) \tag{10.5}$$

$$y(t) = C(t)x(t) + D(t)u(t) \tag{10.6}$$

in which often $D(t) \equiv 0$. A *linear constant* (or *time-invariant*) *differential system* obeys equations such as

$$\dot{x}(t) = Ax(t) + Bu(t) \tag{10.7}$$

$$y(t) = Cx(t) + D_0u(t) + D_1\dot{u}_1(t) + \cdots + D_qu^{(q)}(t) \tag{10.8}$$

where $A, B, C, D_0, D_1, \ldots, D_q$ are constant matrices and often $D_0 = D_1 = \cdots = D_q = 0$.

The system equations (10.5), (10.6) and (10.7), (10.8) are in the particularly simple form in which the differential equations are all first order, and give $\dot{x}(t)$ directly. More complicated equations may arise, such as

$$Q\dot{x}(t) = Ax(t) + Bu(t) \tag{10.9}$$

or

$$Q_1\dot{x}(t) + Q_2\ddot{x}(t) = Ax(t) + Bu(t) \tag{10.10}$$

These are considered in later chapters.

We define *linear time-dependent discrete-time systems* in an analogous way. A typical set of equations for such a system would be

$$x_{k+1} = A_k x_k + B_k u_k \tag{10.11}$$

$$y_k = C_k x_k + D_k u_k \tag{10.12}$$

If A, B, C, D in (10.11) and (10.12) are the same for all k, the system is *constant* (or *time-invariant*). Again the equations may be given in a less simple form.

In engineering applications the matrices occurring in (10.5) to (10.12) have as their entries real numbers or real-valued functions of time. Almost all of the following theory can be developed in terms of real variables: the exceptions arise, for example, when we wish to obtain eigenvalues.

Sometimes, therefore, we consider systems in which the matrices (and therefore x) are complex, but u and y remain always real. It is also sometimes important to know that a transformation which has been developed for complex matrices can also be carried through without departing from the real variables (compare Chapter 2, Theorem 4.1). More general fields than the real or complex numbers could be considered, but some changes would be necessary in proofs, and there seems to be no practical demand for such an extension.

For a linear differential system, a particularly important quantity is the *step response*. This is a real matrix-valued function H of the real variables t, t_0 such that if $x(t_0) = 0$, and $u(t) = e_i U(t - t_0)$, then $y(t) = H(t, t_0)e_i$. Here e_i is the ith column of I_l, and we may regard $H(t, t_0)$ as the output corresponding to a matrix input $I_l U(t - t_0)$. When the system is constant, H becomes a function of the single variable $t - t_0$, $H(t - t_0)$.

The partial derivative $h(t, t_0) = -\partial H(t, t_0)/\partial t_0$ is called the *impulse response*. It may exist only in the sense of the theory of distributions (R & S, pp. 217–223). When the system is time-invariant $h(t, t_0)$ becomes $h(t - t_0)$. Its Laplace transform is then $G(s)$, the *transfer function matrix* (R & S, p. 341).

Theorem 10.1 Let a system with input u and output y satisfy the following conditions:

(i) It is linear. That is, if the response to u_1 is y_1 and the response to u_2 is y_2, then the response to $\alpha_1 u_1 + \alpha_2 u_2$ is $\alpha_1 y_1 + \alpha_2 y_2$ for any real α_1, α_2.

(ii) It is non-anticipative. That is, if $u(t) = 0$, $t \leqslant 0$, then $y(t) = 0$, $t \leqslant 0$.

(iii) If the input u satisfies $u(t) = 0$, $t \leqslant 0$; and for any finite T, $\|u(t)\| < M_1(T)$, $0 < t \leqslant T$; then the output satisfies $\|y(t)\| < M_2(M_1, T)$, $0 < t \leqslant T$.

(iv) The response $H(t_1, t_0)$ at any finite t_1 to the input $I_l U(t - t_0)$ with $t_0 \geqslant 0$ is in D_L^1 as a function of t_0.

Then the response y to an input u satisfying $u(t) = 0$, $t \leqslant 0$; $u \in D_R^1$; is for any finite t

$$y(t) = \int_0^t H(t, \tau)\, du(\tau) \tag{10.13}$$

Proof The proof is an easy adaptation of the proof for time-invariant systems (R & S, pp. 205–208 and Exercise 5.2, pp. 208–209). Let all elements of u^o be zero except u_j^o, and let $|u_j^o(t)| < M_1$ for $0 < t \leqslant T$. Then by (iii),

$$|y_i^o(t)| \leqslant \|y^o(t)\| < M_2, \quad 0 < t \leqslant T, \quad i = 1, 2, \ldots, m \tag{10.14}$$

Now let all elements $u_k(t)$, $k \neq j$, of $u(t)$ be zero for all t, and $u_j(t) = 0$, $0 \leqslant t \leqslant t_1$. It follows from (i) that $y_i(t) = 0, 0 \leqslant t \leqslant t_1$, $i = 1, 2, \ldots m$, for we may put $T = t_1$ in (10.14), and then

$$u_j(t) = 0 = 0 . u_j^o(t), \quad 0 < t \leqslant t_1 \tag{10.15}$$

and so

$$0 \leqslant |y_i(t)| = |0 . y_i^o(t)| \leqslant 0 . M_2 = 0, \quad 0 < t \leqslant t_1, \quad i = 1, 2, \ldots, m \tag{10.16}$$

We also know from (ii) that if $u(t) = 0$, $t \leqslant 0$, then $y_i(t) = 0$, $t \leqslant 0$.

Now consider the integral

$$J = \int_0^{t_1} h_{ij}(t_1, \tau)\, dU(\tau - t_0) \tag{10.17}$$

where h_{ij} is an element of H. The integral exists because $h_{ij}(t_1, \tau) \in D_L^1$ as a function of τ, and $U(\tau - t_0) \in D_R^1$ as a function of τ. By the formula for integration by parts,

$$J = h_{ij}(t_1, t_1)U(t_1 - t_0) - h_{ij}(t_1, 0)U(-t_0) - \int_0^{t_1} U(\tau - t_0)\, dh_{ij}(t_1, \tau) \tag{10.18}$$

But $h_{ij}(t_1, t_1)$ is the response $y_i(t_1)$ to an input $u_j(t) = U(t - t_1)$, and by (10.16) this is zero. Also $U(-t_0) = 0$ if $t_0 \geqslant 0$. Hence (10.18) gives

$$J = -h_{ij}(t_1, t_1) + h_{ij}(t_1, t_0) \tag{10.19}$$

$$= h_{ij}(t_1, t_0) \tag{10.20}$$

Let $u_j \in D_R^1$ be given, with $u_j(t) = 0, t \leqslant 0$. As in R & S, pp. 207–208 we may find a sequence v^n

$$v^n(t) = \sum_{k=1}^{r_n} v^{n,k} U(t - c_i), \quad 0 \leqslant c_i \leqslant T \tag{10.21}$$

such that given any $\delta > 0$ we may find an n_0 with the property that

$$|u_j(t) - v^n(t)| < \delta, \quad 0 \leqslant t \leqslant T, \quad \text{all } n > n_0 \tag{10.22}$$

By (i) the response at $t_1 \in [0, T]$ to an input with jth element v^n is

$$y_i^n(t_1) = \sum_{k=1}^{r_n} v^{n,k} h_{ij}(t_1, c_i) \tag{10.23}$$

which by (10.20) can be written

$$y_i^n(t_1) = \int_0^{t_1} h_{ij}(t_1, \tau)\, dv^n(\tau) \tag{10.24}$$

Also the integral

$$\int_0^{t_1} h_{ij}(t_1, \tau)\, du_j(\tau) \tag{10.25}$$

exists because $h_{ij}(t_1, \tau) \in D_L^1$ as a function of τ, and $u_j(\tau) \in D_R^1$. Then for $0 \leqslant t_1 \leqslant T$, integrating by parts,

$$\left| \int_0^{t_1} h_{ij}(t_1, \tau)\, d[v^n(\tau) - u_j(\tau)] \right| = \left| \int_0^{t_1} [v_n(\tau) - u_j(\tau)]\, dh_{ij}(t_1, \tau) \right| \tag{10.26}$$

$$< M\delta \tag{10.27}$$

where M is a bound for the variation of $h_{ij}(t_1, \tau)$ as a function of τ on $0 \leqslant \tau \leqslant t_1$, which exists because $h(t_1, \tau) \in D_L^1$ as a function of τ. Then (10.27) and (10.24) give

$$\left| y_i^n(t_1) - \int_0^{t_1} h_{ij}(t_1, \tau)\, du_j(\tau) \right| < M\delta \tag{10.28}$$

whenever $n > n_0$.

Also the response y_i^δ of the system to the input $v^n - u_j$ satisfies, by (i) and (iii),

$$|y_i^\delta(t_1)| = |y_i^n(t_1) - y_i(t_1)| < \frac{M_2 \delta}{M_1} \tag{10.29}$$

where y_i is the response to the given u_j. Then (10.28) and (10.29) show that

$$y_i(t) = \int_0^{t_1} h_{ij}(t_1, \tau)\, du_j(\tau) \tag{10.30}$$

which holds for all $t_1 \in [0, T]$, where $T > 0$ can be chosen arbitrarily. Finally by (i) we may superimpose the responses to u_j, $j = 1, 2, \ldots, l$, to give (10.13).

Corollary In the enunciation of the theorem we may substitute $t = \theta$ throughout for $t = 0$.

Proof This amounts to making a change of variable from t to $t - \theta$.

11 Functions

A discussion of functions belongs properly at the beginning of this chapter. We wish here, however, to discuss notation in the light of the requirements of Sections 1 to 10.

A *function* (or *mapping*) f associates, with each member x of a subset D of a set X, a unique member y of a subset R of a set Y. The set D is called the *domain* of f, and the set R is the *range* of f. The y corresponding to an x is called the *image* of x under f, and is written $y = f(x)$. To specify a function completely the following information is needed.

(i) The set X.
(ii) The domain $D \subseteq X$.
(iii) The set Y.
(iv) The range $R \subseteq Y$.
(v) If the sets X and Y are not indicated as such by the notation, the direction of the mapping must be given. For although corresponding to any x there is only one y, corresponding to one y may be more than one x. The information given by the mapping is therefore not symmetric with respect to X and Y.
(vi) A list of each x and the corresponding y, or (more usually) a rule for obtaining y given x.

No existing notation is fully satisfactory in all circumstances. The chief notations in use are the following:

(a) The older tradition was to indicate (v) and (vi) by giving a typical image, for example $y = \sin x$. The sets X, D, Y, R, were specified separately, but as they often remained the same for any one symbol (for example x might always represent any real number) this offered little difficulty. The chief disadvantage of this notation is that it does not easily allow the distinction to be made between f (which is the abstract relation between members of the sets X and Y) and $f(x)$ (which is a member of the set Y). Because this distinction is essential in some branches of analysis, a number of ways of indicating the distinction have arisen.

(α) The notation $f(x) \equiv 0$ means that every $x \in D$ maps into $0 \in Y$, and is therefore the same as saying that $f = 0$ (the zero function).
(β) The notation $f(\cdot)$ is used to mean f in distinction to $f(x)$.

(b) The more modern practice is to indicate a function from X to Y by $f: X \to Y$. An alternative notation is $f: x \to f(x)$, and sometimes a different symbol is used in this context, $f: x \mapsto f(x)$. This notation preserves the distinction between f and $f(x)$ very clearly, but can become tedious or cumbersome in some applications. Part of the required information (in particular D and R) must still be given separately.

A few examples will illustrate these points.

Example 11.1 The 'function sin x' [notation (a)] can be indicated in the following ways when x is real:

$$\sin(\cdot) \tag{11.1}$$

$$f: x \mapsto \sin x \tag{11.2}$$

$$f: \mathscr{R} \to \mathscr{R} \tag{11.3}$$

Each of these gives part of the total information. The domain is $\mathscr{R}$, the set of real numbers, and the range is $[-1, 1] \subset \mathscr{R}$.

Example 11.2 In notation (a) we speak of the 'partial derivative with respect to t_0 of the step response (function) $H(t, t_0)$'. This can be written

$$\frac{\partial}{\partial t_0} H(t, t_0) \quad \text{[notation (a)]} \tag{11.4}$$

$$D_2H(.,.) \quad \left[\text{but not } \frac{\partial}{\partial \cdot} H(\cdot,\cdot)\right] \tag{11.5}$$

$$f: (t, t_0) \mapsto \frac{\partial}{\partial t_0} H(t, t_0) \tag{11.6}$$

$$f: \mathscr{R} \times \mathscr{R} \to \mathscr{R} \tag{11.7}$$

Example 11.3 The Laplace transform of a function f is another function $\bar{f}$. We may write the Laplace transformation $\mathscr{L}$ in the forms

$$\mathscr{L}\{f\} \quad \text{[notation (a)]} \tag{11.8}$$

$$\mathscr{L}\{\cdot\} \tag{11.9}$$

$$\mathscr{L}: f \mapsto \mathscr{L}\{f\} \tag{11.10}$$

$$\mathscr{L}: X \to Y \tag{11.11}$$

where X is a function space with elements f, and Y is a function space with elements $\bar{f}$. Notice that $\mathscr{L}\{f\}$ is the image of f under $\mathscr{L}$, that is $\bar{f}$. It is sometimes convenient to write this image

$$\mathscr{L}\{f(t)\} = \bar{f}(s) \tag{11.12}$$

where $f(t)$ represents f and $\bar{f}(s)$ represents $\bar{f}$ [notation (a)]. For example we write

$$\mathscr{L}\{e^{-at}\} = \frac{1}{s+a} \tag{11.13}$$

Example 11.4 The 'unit step function at t_0' is defined by

$$f(t) = \begin{cases} 0, & t \leqslant t_0 \\ 1, & t > t_0 \end{cases} \tag{11.14}$$

and is usually written $U(t - t_0)$. We cannot write U or $U(\cdot)$ as we then lose t_0. The notation U_{t_0} or $U_{t_0}(t)$ is unsatisfactory because it fails to show that U is a function only of $t - t_0$.

In this book, no confusion is likely to arise between a function and an image, and the older notation (a) would be satisfactory and generally convenient. In deference to the importance of the distinction in other areas, however, functions have generally been indicated by f (for example) and images by $f(x)$. Minor inconveniences arising from this usage have been tolerated, but major inconveniences have been avoided by reverting to the older convention. It is hoped that the meaning will always be clear.

In order to relate functions of time (say u) to their Laplace transforms, the same letter is used for the transform with a bar above it (thus $\bar{u}$). Similarly z-transforms are denoted by a tilde (thus $\tilde{u}$). Where it is possible to omit the bar or tilde without confusion, this is done. For example we write G for the transfer function matrix instead of $\bar{G}$, because the corresponding time function is traditionally represented by h. With exceptions due to tradition (such as h) matrices are represented by capital letters.

2 Linear constant differential systems

In this chapter we shall be concerned with linear systems which are constant (that is, time-invariant) and which are described by sets of ordinary differential equations.

1 State-space equations

Linear constant systems described by differential equations can be treated by means of the Laplace transformation (Chapter 1, Section 9). Let the input to the system be an l-vector u which as a function of time t belongs to D_R (Chapter 1, Section 5), is zero for $t \leqslant 0$, and satisfies

$$|u_i(t)| < Me^{\alpha t}, \quad t > 0, \quad i = 1, 2, \ldots, l \tag{1.1}$$

Then u has a Laplace transform $\bar{u}$ valid for $\operatorname{Re} s > \alpha$.

Suppose first, as the simplest case, that u is continuous and the equations of the system are

$$\dot{x} = Ax + Bu \tag{1.2}$$

$$y = Cx \tag{1.3}$$

where x is an n-vector (the state) and y is an m-vector (the output). The matrices A, B, and C are constant and are respectively $n \times n$, $n \times l$, and $m \times n$; they will be called respectively the *plant matrix*, the *input matrix*, and the *output matrix*. System equations such as (1.2) and (1.3) will be said to be in *state-space form*. The order of the system is n, which is the dimension of the state vector and is the number of independent initial conditions required by equation (1.2).

If the initial condition for x in (1.2) is $x(0) = 0$, then on taking Laplace transforms the result is

$$s\bar{x} = A\bar{x} + B\bar{u} \tag{1.4}$$

$$\bar{x} = (sI_n - A)^{-1}B\bar{u} \tag{1.5}$$

and

$$\bar{y} = C\bar{x} = C(sI_n - A)^{-1}B\bar{u} \tag{1.6}$$

The $m \times l$ rational matrix $C(sI_n - A)^{-1}B$ summarizes the response of the system so far as its external behaviour is concerned. This matrix is called the *transfer function matrix* and is given the symbol $G(s)$.

Theorem 1.1 Write

$$|sI_n - A| = s^n + a_{n-1}s^{n-1} + \cdots + a_1 s + a_0 \tag{1.7}$$

Then

$$\begin{aligned} G(s) = \{(s^{n-1} + a_{n-1}s^{n-2} + \cdots + a_1)CB \\ + (s^{n-2} + a_{n-1}s^{n-3} + \cdots + a_2)CAB \\ + \cdots + (s + a_{n-1})CA^{n-2}B + CA^{n-1}B\} \\ \div (s^n + a_{n-1}s^{n-1} + \cdots + a_0) \end{aligned} \tag{1.8}$$

Proof This follows at once from equation (1.6) and Chapter 1, Theorem 2.2.

Theorem 1.2 The transfer function matrix $G(s)$ is given by

$$G(s) = C\Gamma(s)B/m(s) \tag{1.9}$$

where $\Gamma(s)$ is the reduced adjoint matrix of $sI_n - A$ and $m(s)$ is the minimal polynomial of A.

Proof See Chapter 1, Section 2.

If $G(s)$ is written $N(s)/d(s)$, where $N(s)$ is a polynomial matrix and $d(s)$ is a monic polynomial having no factor in common with every element of $N(s)$, it should not be concluded from Theorem 1.2 that $d(s)$ is the minimal polynomial of A. For although $m(s)$ has no factor in common with every element of $\Gamma(s)$, it may have factors in common with every element of $C\Gamma(s)B$ (Chapter 1, Section 2). For conditions which prevent this, see Corollary 3 of Theorem 4.2 in Chapter 3.

Example 1.1 Figure 1 represents two masses m_1 and m_2 connected to springs with rates r_1 and r_2. The displacements of the masses are x_1 and x_2: define $x_3 = \dot{x}_1$, $x_4 = \dot{x}_2$ so that x_3 and x_4 are the velocities of the two masses. The input to the system is a force u applied to the lower mass, and the output is $y = x_2$. Then the system obeys the equations

$$\left.\begin{aligned} \dot{x}_1 &= x_3 \\ \dot{x}_2 &= x_4 \\ \dot{x}_3 &= [u - r_1(x_1 - x_2)]/m_1 \\ \dot{x}_4 &= [r_1(x_1 - x_2) - r_2 x_2]/m_2 \end{aligned}\right\} \tag{1.10}$$

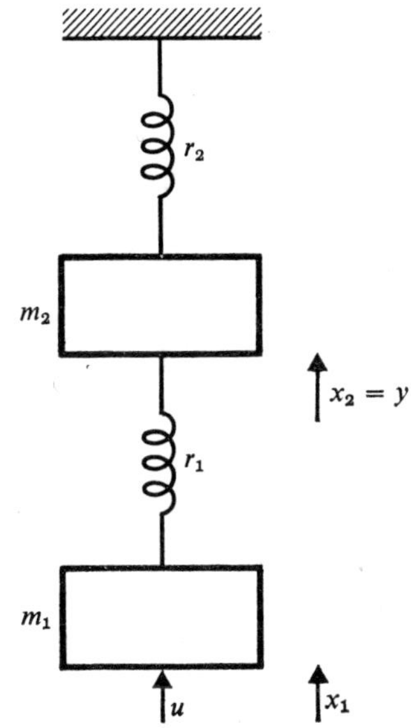

FIG. 1

If u obeys all the conditions given earlier, and if the initial conditions are $x_1(0) = x_2(0) = x_3(0) = x_4(0) = 0$ then equations (1.10) can be Laplace transformed to give

$$s\begin{pmatrix}\bar{x}_1\\ \bar{x}_2\\ \bar{x}_3\\ \bar{x}_4\end{pmatrix} = \begin{pmatrix}0 & 0 & 1 & 0\\ 0 & 0 & 0 & 1\\ -r_1/m_1 & r_1/m_1 & 0 & 0\\ r_1/m_2 & -(r_1+r_2)/m_2 & 0 & 0\end{pmatrix}\begin{pmatrix}\bar{x}_1\\ \bar{x}_2\\ \bar{x}_3\\ \bar{x}_4\end{pmatrix} + \begin{pmatrix}0\\ 0\\ 1/m_1\\ 0\end{pmatrix}\bar{u} \quad (1.11)$$

$$\bar{y} = (0 \quad 1 \quad 0 \quad 0)\begin{pmatrix}\bar{x}_1\\ \bar{x}_2\\ \bar{x}_3\\ \bar{x}_4\end{pmatrix} \quad (1.12)$$

where the matrices A, B, and C are shown explicitly. The transfer function matrix (here 1×1) is

$$G(s) = (0 \quad 1 \quad 0 \quad 0)\begin{pmatrix}s & 0 & -1 & 1\\ 0 & s & 0 & -1\\ r_1/m_1 & -r_1/m_1 & s & 0\\ -r_1/m_2 & (r_1+r_2)/m_2 & 0 & s\end{pmatrix}^{-1}\begin{pmatrix}0\\ 0\\ 1/m_1\\ 0\end{pmatrix} \quad (1.13)$$

$$= -\frac{1}{m_1}\begin{vmatrix} s & -1 & 0 \\ 0 & 0 & -1 \\ -r_1/m_2 & 0 & s \end{vmatrix} \div \begin{vmatrix} s & 0 & -1 & 0 \\ 0 & s & 0 & -1 \\ r_1/m_1 & -r_1/m_1 & s & 0 \\ -r_1/m_2 & (r_1 + r_2)/m_2 & 0 & s \end{vmatrix} \tag{1.14}$$

by Cramer's rule, whence

$$G(s) = \frac{r_1}{m_1 m_2 s^4 + [m_2 r_1 + m_1(r_1 + r_2)]s^2 + r_1 r_2} \tag{1.15}$$

EXERCISE 1.1 In Example 1.1 the state vector could have been defined in many different ways; for example by writing $x_3 = m_1\dot{x}_1$, $x_4 = m_2\dot{x}_2$. Investigate this and some similar changes in equations (1.10) and show that they leave $G(s)$ unchanged.

EXERCISE 1.2 Verify equation (1.8) for the system in Example 1.1.

Example 1.2 The electrical circuit in Fig. 2 obeys the equations

$$\left.\begin{aligned} e_1 &= L_1 \frac{d}{dt}(i - i_1) \\ e_1 &= R_1 i_1 \\ e_2 &= R_2 i \\ e - e_1 - e_2 &= L_2 \frac{di}{dt} \end{aligned}\right\} \tag{1.16}$$

where e is regarded as the input and i as the output. Substitute from the second and third of equations (1.16) into the fourth to give

$$\frac{di}{dt} = -\frac{R_2}{L_2} i - \frac{R_1}{L_2} i_1 + \frac{1}{L_2} e \tag{1.17}$$

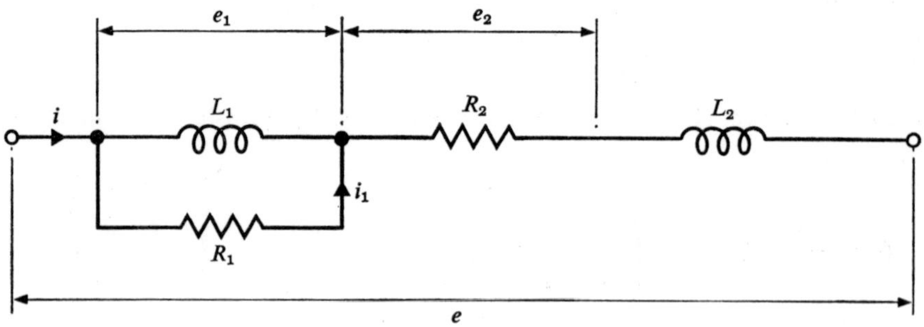

FIG. 2

Then substitute from equation (1.17) and the second of equations (1.16) into the first of equations (1.16) to give

$$\frac{di_1}{dt} = -\frac{R_2}{L_2}i - R_1\left(\frac{1}{L_1} + \frac{1}{L_2}\right)i_1 + \frac{1}{L_2}e \tag{1.18}$$

This shows that $\begin{pmatrix} i \\ i_1 \end{pmatrix}$ is a state vector, and the matrices A, B, and C are

$$A = \begin{pmatrix} -R_2/L_2 & -R_1/L_2 \\ -R_2/L_2 & -R_1(1/L_1 + 1/L_2) \end{pmatrix} \tag{1.19}$$

$$B = \begin{pmatrix} 1/L_2 \\ 1/L_2 \end{pmatrix} \tag{1.20}$$

$$C = (1 \quad 0) \tag{1.21}$$

The transfer function is now the admittance $Y(s) = \bar{i}(s)/\bar{e}(s)$ of the circuit and is

$$(1 \quad 0)\begin{pmatrix} s + R_2/L_2 & R_1/L_2 \\ R_2/L_2 & s + R_1(1/L_1 + 1/L_2) \end{pmatrix}^{-1}\begin{pmatrix} 1/L_2 \\ 1/L_2 \end{pmatrix} \tag{1.22}$$

$$= \frac{L_1 s + R_1}{L_1 L_2 s^2 + [R_1(L_1 + L_2) + R_2 L_1]s + R_1 R_2} \tag{1.23}$$

which can easily be confirmed by direct calculation. This example shows that the transfer function matrix may represent driving point immittances, as well as transfer immittances. It also shows that without some systematic procedure it may not be obvious how to set up the state-space equations for a system.

EXERCISE 1.3 By defining a current $i_2 = i - i_1$ obtain a different set of state-space equations for the circuit in Fig. 2. Show that the same admittance is obtained.

If u is not continuous, but has simple discontinuities on the right as permitted in the class D_R^1, then equation (1.2) is supposed to hold only where u is continuous. At points of discontinuity of u, the solution x of (1.2) is continuous. The Laplace transformation can still be applied without difficulty to equations (1.2) and (1.3) (Chapter 1, Theorem 9.2).

Example 1.3 In Example 1.1, suppose that the force u is a unit step at $t = 0$,

$$u(t) = U(t) \tag{1.24}$$

Then

$$\bar{u}(s) = \frac{1}{s} \tag{1.25}$$

and

$$\bar{y}(s) = \bar{x}_2(s) = G(s)\bar{u}(s)$$
$$= r_1/s\{m_1m_2s^4 + [m_2r_1 + m_1(r_1 + r_2)]s^2 + r_1r_2\} \qquad (1.26)$$

EXERCISE 1.4 Put the last expression in (1.26) into partial fractions and so find y.

For some purposes it is desirable to augment equation (1.3) to allow the output y to depend not only on the state x but also on the input u and its derivatives. This leads, after Laplace transformation, to the equation

$$\bar{y} = C\bar{x} + D(s)\bar{u} \qquad (1.27)$$

in place of (1.6), where $D(s)$ is a polynomial matrix. It is assumed, when (1.27) is written, that u has derivatives of the appropriate order. The sense in which these derivatives exist can, however, be given different interpretations. Some of the possibilities are as follows:

(i) The derivatives of u may be assumed to exist everywhere in the usual sense of analysis.
(ii) The derivatives of u may be assumed to exist everywhere except at a finite number of points on any finite interval: at the exceptional points both left-hand and right-hand derivatives exist but they are unequal.
(iii) The derivatives of u exist in the distribution sense.

Of these possibilities the second will usually be adopted in this book (as in R & S).

Example 1.4 Let a system have one input u and one output y, and let $D(s) = s$, so that the contribution to y is du/dt. If

$$u(t) = (t - 1)^2U(t - 1) \qquad (1.28)$$

then u has a derivative everywhere in the usual sense, equal to $2(t - 1)U(t - 1)$ If

$$u(t) = (t - 1)U(t - 1) \qquad (1.29)$$

then u has a left-hand derivative and a right-hand derivative at every point, and these are equal except at $t = 1$. If

$$u(t) = U(t - 1) \qquad (1.30)$$

then u does not have a derivative in the usual sense at $t = 1$, but in the distribution sense the derivative of u is $\delta(t - 1)$.

1.1 System matrix

Equations (1.4) and (1.27) can be written as a single equation,

$$\begin{pmatrix} sI_n - A & B \\ -C & D(s) \end{pmatrix} \begin{pmatrix} \bar{x} \\ -\bar{u} \end{pmatrix} = \begin{pmatrix} 0 \\ -\bar{y} \end{pmatrix} \tag{1.31}$$

The $(n + m) \times (n + l)$ polynomial matrix

$$P(s) = \begin{pmatrix} sI_n - A & B \\ -C & D(s) \end{pmatrix} \tag{1.32}$$

which occurs in (1.31) contains all the mathematical information about the system which is needed to discuss its behaviour. Accordingly $P(s)$ is called the *system matrix*, and because it is in a special form corresponding to the state-space equations, $P(s)$ is said to be in *state-space* form (in earlier papers the term *first form* was used). In general, any description which may be applied to a system will be applied also to its system matrix.

Write C in terms of its rows and B in terms of its columns

$$C = \begin{pmatrix} c_1 \\ c_2 \\ \vdots \\ c_m \end{pmatrix}, \qquad B = (b_1\, b_2 \ldots b_l) \tag{1.33}$$

Then from the equation

$$G(s) = C(sI_n - A)^{-1}B + D(s) \tag{1.34}$$

which follows from (1.31), it is clear that

$$g_{ij}(s) = c_i(sI_n - A)^{-1}b_j + d_{ij}(s) \tag{1.35}$$

where as usual g_{ij}, d_{ij} are elements of $G = (g_{ij})$, $D = (d_{ij})$. By equation (1.24) of Chapter 1 it follows that

$$g_{ij}(s) = \frac{P^{1,2,\ldots,n,n+i}_{1,2,\ldots,n,n+j}}{|sI_n - A|} \tag{1.36}$$

where $P^{1,2,\ldots,n,n+i}_{1,2,\ldots,n,n+j}$ is the minor formed from rows $1, 2, \ldots, n, n + i$ and columns $1, 2, \ldots, n, n + j$ of $P(s)$. In exactly the same way any minor of $G(s)$ can be written in terms of $P(s)$,

$$G^{i_1,i_2,\ldots,i_q}_{j_1,j_2,\ldots,j_q} = \frac{P^{1,2,\ldots,n,n+i_1,n+i_2,\ldots,n+i_q}_{1,2,\ldots,n,n+j_1,n+j_2,\ldots,n+j_q}}{|sI_n - A|} \tag{1.37}$$

It is convenient to abbreviate the indices occurring on P in (1.36) and (1.37) and to write

$$P^{1,2,\ldots,n,n+i_1,n+i_2,\ldots,n+i_q}_{1,2,\ldots,n,n+j_1,n+j_2,\ldots,n+j_q} = P^{i_1,i_2,\ldots,i_q}_{j_1,j_2,\ldots,j_q}) \tag{1.38}$$

With this notation

$$P) = |sI_n - A| \tag{1.39}$$

and we may state the following theorem.

Theorem 1.3 The minors of G are given by

$$G^{i_1,i_2,\ldots,i_q}_{j_1,j_2,\ldots,j_q} = P^{i_1,i_2,\ldots,i_q}_{j_1,j_2,\ldots,j_q}) \div P) \tag{1.40}$$

Example 1.5 In Example 1.1 the system matrix is

$$P(s) = \left(\begin{array}{cccc|c} s & 0 & -1 & 0 & 0 \\ 0 & s & 0 & -1 & 0 \\ r_1/m_1 & -r_1/m_1 & s & 0 & 1/m_1 \\ -r_1/m_2 & (r_1+r_2)/m_2 & 0 & s & 0 \\ \hline 0 & -1 & 0 & 0 & 0 \end{array}\right) \tag{1.41}$$

and equation (1.40) gives

$$G = g_{11} = \begin{vmatrix} s & 0 & -1 & 0 & 0 \\ 0 & s & 0 & -1 & 0 \\ r_1/m_1 & -r_1/m_1 & s & 0 & 1/m_1 \\ -r_1/m_2 & (r_1+r_2)/m_2 & 0 & s & 0 \\ 0 & -1 & 0 & 0 & 0 \end{vmatrix}$$

$$\div \begin{vmatrix} s & 0 & -1 & 0 \\ 0 & s & 0 & -1 \\ r_1/m_1 & -r_1/m_1 & s & 0 \\ -r_1/m_2 & (r_1+r_2)/m_2 & 0 & s \end{vmatrix} \tag{1.42}$$

On expanding the numerator determinant by the last row and last column, this will be found to agree with equation (1.14).

This example shows that equation (1.40) will not usually be a profitable formula for numerical evaluation. For theoretical purposes, however, the formula is of fundamental importance.

Theorem 1.4 If in the system matrix $P(s)$ of equation (1.32), $D(s) = 0$, then no $q \times q$ minor of $G(s)$ has its numerator of higher degree in s than $n - q$.

Proof Evaluate the minor in the numerator on the right-hand side of (1.40) by its Laplace expansion in terms of the $q \times q$ minors formed from the last q columns. These $q \times q$ minors are independent of s, and the complementary minors have degree at most $n - q$.

Corollary If $G(s) \to 0$ as $s \to \infty$ (G is then said to be *proper*) every $q \times q$ minor of $G(s)$ tends to zero at least as fast as s^{-q}.

Proof $G(s) \to 0$ as $s \to \infty$ if and only if $D(s) = 0$. It then follows from Theorem 1.4 and equation (1.40) that $s^q G^{i_1, i_2, \ldots, i_q}_{j_1, j_2, \ldots, j_q}$ tends to a finite limit as $s \to \infty$.

EXERCISE 1.5 Prove the corollary by direct expansion of the $q \times q$ minors of G.

Theorem 1.5 If $l = m$ and $|G(s)| \not\equiv 0$, the minors of $Q = G^{-1}$ are given by

$$Q^{i_1, i_2, \ldots, i_k}_{j_1, j_2, \ldots, j_k} = \frac{(-1)^{i_1+i_2+\cdots+i_k+j_1+j_2+\cdots+j_k} P^{j'_1, j'_2, \ldots, j'_{m-k}}_{i'_1, i'_2, \ldots, i'_{m-k}}}{P^{1, 2, \ldots, m}_{1, 2, \ldots, m}} \tag{1.43}$$

where $i_1 < i_2 < \cdots < i_k$ and $i'_1 < i'_2 < \cdots < i'_{m-k}$ together form a complete set of indices $1, 2, \ldots, m$, and similarly $j_1 < j_2 < \cdots < j_k$ and $j'_1 < j'_2 < \cdots < j'_{m-k}$ form a complete set of indices. In particular

$$q_{ij} = Q^i_j = \frac{(-1)^{i+j} P^{1, 2, \ldots, j-1, j+1, \ldots, m}_{1, 2, \ldots, i-1, i+1, \ldots, m}}{P^{1, 2, \ldots, m}_{1, 2, \ldots, m}} \tag{1.44}$$

Proof This follows from (1.40) and the formula for minors of the inverse matrix (Chapter 1, Theorem 1.4).

2 General linear constant differential equations

It was seen in Section 1 that if the equations of a physical system are written down, they may not initially be in state-space form. For linear mechanical or electrical systems there are systematic procedures which will produce state-space equations. Often, however, the physical system is nonlinear, and linear equations are obtained by considering small perturbations from a steady state.

In such a case the resulting equations will not usually be in state-space form, and it may not be obvious how they can be brought to this form. For example, the equations which result from linearization may be mixed algebraic and differential equations. Alternatively they may be in the form

$$Q\dot{x} = Ax + Bu \tag{2.1}$$

with Q singular; or the order of some of the differential equations may be greater than 1.

For several reasons, therefore, it is desirable to have a more general way of describing a system. It will be assumed that the system satisfies linear algebraic and differential equations with constant coefficients, and that the input u satisfies appropriate conditions as before. After taking the Laplace transform with zero initial conditions let the result be

$$T(s)\bar{\xi} = U(s)\bar{u} \tag{2.2}$$

$$\bar{y} = V(s)\bar{\xi} + W(s)\bar{u} \tag{2.3}$$

where T, U, V, W are all polynomial matrices, respectively $r \times r$, $r \times l$, $m \times r$, $m \times l$. We demand that the determinant of $T(s)$ should not be identically zero, for if it were equation (2.2) would be indeterminate. Notice that the symbol U is used also for the unit step $U(t)$, but this is always written with its argument [which is indeed necessary: cf. $U(t - 1)$] so that no confusion can arise.

Example 2.1 From Fig. 1,

$$\left.\begin{aligned} m_1\ddot{x}_1 &= u - r_1(x_1 - x_2) \\ m_2\ddot{x}_2 &= r_1(x_1 - x_2) - r_2x_2 \end{aligned}\right\} \tag{2.4}$$

so that here

$$\bar{\xi} = \begin{pmatrix} \bar{x}_1 \\ \bar{x}_2 \end{pmatrix}, \quad T(s) = \begin{pmatrix} m_1s^2 + r_1 & -r_1 \\ -r_1 & m_2s^2 + r_1 + r_2 \end{pmatrix}, \quad U(s) = \begin{pmatrix} 1 \\ 0 \end{pmatrix},$$

$$V(s) = (0 \quad 1), \qquad W(s) = 0 \tag{2.5}$$

Example 2.2 On writing

$$\xi = \begin{pmatrix} i \\ i_1 \\ e_1 \\ e_2 \end{pmatrix}, \qquad u = e, \qquad y = i \tag{2.6}$$

equations (1.16) become, after Laplace transformation,

$$\begin{pmatrix} L_1s & -L_1s & -1 & 0 \\ 0 & R_1 & -1 & 0 \\ R_2 & 0 & 0 & -1 \\ L_2s & 0 & 1 & 1 \end{pmatrix} \bar{\xi} = \begin{pmatrix} 0 \\ 0 \\ 0 \\ 1 \end{pmatrix} \bar{u} \tag{2.7}$$

$$\bar{y} = (1 \quad 0 \quad 0 \quad 0)\bar{\xi} \tag{2.8}$$

where (2.7) and (2.8) show T, U, and V explicitly.

Example 2.3 The equations

$$\left.\begin{aligned} \frac{d\xi_1}{dt} + \frac{d^3\xi_2}{dt^3} &= -\xi_1 \\ \frac{d\xi_2}{dt} &= -\xi_2 + u \\ y &= \xi_1 \end{aligned}\right\} \tag{2.9}$$

together with the initial conditions

$$\xi_1(0) = \xi_2(0) = \dot{\xi}_2(0) = \ddot{\xi}_2(0) = 0 \tag{2.10}$$

give rise to

$$T(s) = \begin{pmatrix} s+1 & s^3 \\ 0 & s+1 \end{pmatrix}, \qquad U(s) = \begin{pmatrix} 0 \\ 1 \end{pmatrix}, \qquad V(s) = (1 \quad 0) \tag{2.11}$$

Having generalized the way of writing our differential equations, it is now necessary to say what is the order of the system. One way of doing this is to make the appropriate definition, develop the theory, and show that the results are consistent with the definition of order for state-space equations. The appropriate definition for this purpose is that the *order* of the system represented by equations (2.2) and (2.3) is the degree of the determinant of $T(s)$. A justification of this definition will be given in Section 8 of Chapter 5.

Example 2.4 The orders of the systems in Examples 2.1 and 2.2 are respectively 4 and 2. This agrees with their state-space equations (1.11) and (1.19).

Example 2.5 The order of the system in Example 2.3 is 2. To bring equations (2.9) to state-space form differentiate and change sign in the second equation twice, and then add to obtain

$$\left.\begin{aligned} \frac{d\xi_2}{dt} &= -\xi_2 + u \\ -\frac{d^2\xi_2}{dt^2} &= \frac{d\xi_2}{dt} - \frac{du}{dt} \\ \frac{d^3\xi_2}{dt^3} &= -\frac{d^2\xi_2}{dt^2} + \frac{d^2u}{dt^2} \end{aligned}\right\} \tag{2.12}$$

$$\frac{d^3\xi_2}{dt^3} = -\xi_2 + u - \frac{du}{dt} + \frac{d^2u}{dt^2} \tag{2.13}$$

Substitute from (2.13) into the original equations to give

$$\left.\begin{aligned} \frac{d\xi_1}{dt} &= -\xi_1 + \xi_2 - u + \frac{du}{dt} - \frac{d^2u}{dt^2} \\ \frac{d\xi_2}{dt} &= \qquad\;\; -\xi_2 + u \end{aligned}\right\} \tag{2.14}$$

Now put

$$x_1 = \xi_1 + \frac{du}{dt} - 2u, \qquad x_2 = \xi_2 \tag{2.15}$$

and obtain

$$\left.\begin{aligned} \frac{dx_1}{dt} &= -x_1 + x_2 - 3u \\ \frac{dx_2}{dt} &= \qquad - x_2 + u \\ y &= \quad x_1 - \frac{du}{dt} + 2u \end{aligned}\right\} \tag{2.16}$$

Equations (2.16) are in state-space form and are of second order.

2.1 Justification of the definition of order

Though the suggested way of justifying the definition of order for the system of equations (2.2) and (2.3) is the simplest and neatest, it is not very informative at this stage for the reader. A justification of the definition is therefore given here, but it may be omitted if desired.

Our working definition of the order of a system (Chapter 1, Section 10) is the number of independent initial conditions needed to define the solution of its equations when the input $u = 0$. This is by no means the same as the number of initial conditions needed to apply the Laplace transformation, for these may not be independent. For example, on putting $u = 0$ in the second of equations (2.9),

$$\frac{d\xi_2}{dt} = -\xi_2, \qquad \frac{d^2\xi_2}{dt^2} = -\frac{d\xi_2}{dt} = \xi_2 \tag{2.17}$$

and so the third and fourth of the initial conditions in (2.10) are implied by the second. Only two independent initial conditions are needed, and the system has order 2.

To find how many independent initial conditions will be needed in the differential equation giving rise to (2.2), put $u = 0$ and replace s by the differential operator D. The number of independent parameters in the complementary function, which is the solution of

$$T(D)\xi = 0 \tag{2.18}$$

is the required order of the system. Consider the three following types of elementary operation on the set of simultaneous differential equations represented by (2.18).

(i) Multiply any row by a non-zero constant α.
(ii) Interchange rows i and j.
(iii) Add to row i a multiple by a constant β of the result of differentiating row j any desired number of times.

A solution of the equations remains a solution after any such operations have been carried out. Moreover, each such operation has an inverse of the same type. Consequently any solution of the transformed equations is also a solution of the original equations.

By a sequence of such elementary operations $T(D)$ can be brought to lower triangular form $Q(D)$ (compare Chapter 1, Theorem 1.6). The first row of the equation $Q(D)\xi = 0$ is $q_{11}(D)\xi_1 = 0$. When this is integrated, a number of arbitrary constants will be introduced equal to the degree of $q_{11}(D)$. The second equation is then

$$q_{22}(D)\xi_2 = -q_{21}(D)\xi_1 \tag{2.19}$$

where the right-hand side is now a known function. When equation (2.19) is integrated a further set of arbitrary constants will be introduced, equal in number to the degree of $q_{22}(D)$. It follows in this way that the total number of arbitrary constants in the solution of $Q(D)\xi = 0$ (which is also the solution of $T(D)\xi = 0$) is equal to the sum of the degrees of $q_{11}(D), q_{22}(D), \ldots, q_{rr}(D)$. As Q is triangular, this is the degree of $|Q(D)|$, which is the same as the degree of $|T(D)|$ because the elementary operations leave the degree of the determinant unchanged. With the substitution of s for D this completes the justification of our definition of order.

It should be stressed that in equation (2.2), cancellation is not permitted: or at least if it is carried out the order of the equations is changed. That is to say, the equations

$$(s^2 + 3s + 2)\bar{\xi} = (s + 1)\bar{u} \tag{2.20}$$

and

$$(s + 2)\bar{\xi} = \bar{u} \tag{2.21}$$

must be regarded as distinct equations even though they have the same solution. The two systems described by equations (2.20), (2.21) have order 2 and 1 respectively. Correspondingly, two independent initial conditions are needed to set up (2.20), while only one is needed to set up (2.21).

2.2 Polynomial system matrices

Just as the state-space equations can be assembled into a system matrix, so also can the more general equations (2.2) and (2.3). One preliminary change may be needed, however, for reasons which will become clear in Chapter 3.

In equation (2.2), T is an $r \times r$ matrix, and r need bear no relation to the order n of the system, which is the degree of $|T(s)|$. In Example 2.1, $r = 2, n = 4$; in Example 2.2, $r = 4, n = 2$; while in Example 2.3, $r = n = 2$. Before assembling T, U, V, and W into a system matrix we shall always ensure that $r \geqslant n$. If this is not true of the given matrices, they are trivially expanded to T_1, U_1, V_1, W_1 in the following way

$$T_1 = \begin{pmatrix} I_{n-r} & 0 \\ 0 & T \end{pmatrix}, \qquad U_1 = \begin{pmatrix} O_{n-r,l} \\ U \end{pmatrix}, \qquad V_1 = (0_{m,n-r} \quad V), \qquad W_1 = W \tag{2.22}$$

This amounts to adding new algebraic equations of the form

$$\bar{\xi}' = 0_{n-r,l}\bar{u} \tag{2.23}$$

and writing

$$\bar{y} = 0_{m,n-r}\bar{\xi}' + V(s)\bar{\xi} + W(s)\bar{u} \tag{2.24}$$

With the understanding, therefore, that $r \geqslant n$, the system matrix is defined as

$$P(s) = \begin{pmatrix} T(s) & U(s) \\ -V(s) & W(s) \end{pmatrix} \tag{2.25}$$

This system matrix is said to be in *polynomial form* (elsewhere, *second form*). The polynomial form of the system matrix clearly includes the state-space form as a special case.

From equations (2.2) and (2.3),

$$\bar{\xi} = T^{-1}(s)U(s)\bar{u} \tag{2.26}$$

$$\bar{y} = [V(s)T^{-1}(s)U(s) + W(s)]\bar{u} \tag{2.27}$$

from which the transfer function matrix is found to be

$$G(s) = V(s)T^{-1}(s)U(s) + W(s) \tag{2.28}$$

Let the notation $P^{i_1, i_2, \ldots, i_q}_{j_1, j_2, \ldots, j_q})$ which was introduced in Section 1 be generalized to mean the minor formed from rows $1, 2, \ldots, r, r + i_1, r + i_2 \ldots, r + i_q$ and columns $1, 2, \ldots, r, \ r + j_1, r + j_2, \ldots, r + j_q$ of $P(s)$. Then the following theorem generalizes Theorem 1.3.

Theorem 2.1 The minors of G are given by

$$G^{i_1, i_2, \ldots, i_q}_{j_1, j_2, \ldots, j_q} = \frac{P^{i_1, i_2, \ldots, i_q}_{j_1, j_2, \ldots, j_q})}{P)} \tag{2.29}$$

Proof This follows from equation (2.28) as in the proof of Theorem 1.3.

Example 2.6 From equations (2.5),

$$G(s) = \begin{vmatrix} I_2 & 0 & 0 & 0 \\ 0 & m_1s^2 + r_1 & -r_1 & 1 \\ 0 & -r_1 & m_2s^2 + r_1 + r_2 & 0 \\ 0 & 0 & -1 & 0 \end{vmatrix}$$

$$\div \begin{vmatrix} I_2 & 0 & 0 \\ 0 & m_1s^2 + r_1 & -r_1 \\ 0 & -r_1 & m_2s^2 + r_1 + r_2 \end{vmatrix} \tag{2.30}$$

$$= \frac{r_1}{m_1m_2s^4 + [m_2r_1 + m_1(r_1 + r_2)]s^2 + r_1r_2} \tag{2.31}$$

which is the same as (1.15). The redundant (first and second) rows and columns in P could clearly have been omitted when evaluating equation (2.29).

Exercise 2.1 Use equation (2.29) with equations (2.7) and (2.8) to verify equation (1.23).

Exercise 2.2 Find the transfer function of the system in Example 2.3 from equation (2.28) and equation (2.29), and from equations (2.9) and equations (2.16).

Example 2.7 The system matrix

$$P(s) = \left(\begin{array}{cc:c} 1 & 0 & 0 \\ 0 & (s+1)^2 & s^3 \\ \hdashline 0 & -1 & 2-s \end{array}\right) \tag{2.32}$$

gives

$$G(s) = \frac{s^3}{(s+1)^2} + 2 - s = \frac{3s + 2}{(s+1)^2} \tag{2.33}$$

Notice that the non-vanishing of $W(s)$ does not prevent $G(s)$ from being proper. Indeed, there is no easy way to tell, from a general polynomial system matrix, whether $G(s)$ is proper or not.

Exercise 2.3 Prove the formula corresponding to (1.43) for a polynomial system matrix.

3 Transformation of system matrices

The advantage of using the system matrix $P(s)$ is that all transformations of the system equations can be expressed as operations on $P(s)$. By this means the operations can be systematized and their properties more easily studied. We shall be particularly interested in transformations which leave unchanged the transfer function matrix $G(s)$ and the system order n.

3.1 Strict system equivalence

Let an $(r + m) \times (r + l)$ polynomial system matrix $P(s)$ be given. Let $M(s)$, $N(s)$ be $r \times r$ unimodular polynomial matrices; that is, their determinants are nonzero and independent of s. Also let $X(s)$, $Y(s)$ be polynomial matrices, respectively $m \times r$ and $r \times l$. If

$$P(s) = \begin{pmatrix} T(s) & U(s) \\ -V(s) & W(s) \end{pmatrix} \tag{3.1}$$

and

$$P_1(s) = \begin{pmatrix} T_1(s) & U_1(s) \\ -V_1(s) & W_1(s) \end{pmatrix} \tag{3.2}$$

are related by the transformation

$$\begin{pmatrix} M(s) & 0 \\ X(s) & I_m \end{pmatrix}\begin{pmatrix} T(s) & U(s) \\ -V(s) & W(s) \end{pmatrix}\begin{pmatrix} N(s) & Y(s) \\ 0 & I_l \end{pmatrix} = \begin{pmatrix} T_1(s) & U_1(s) \\ -V_1(s) & W_1(s) \end{pmatrix} \tag{3.3}$$

then we shall say that $P(s)$ and $P_1(s)$ are *strictly system equivalent.*

Theorem 3.1 Two system matrices which are strictly system equivalent have the same order and give rise to the same transfer function matrix.

Proof Equation (3.3) gives

$$T_1(s) = M(s)T(s)N(s) \tag{3.4}$$

On taking determinants and noticing that M and N are unimodular it follows that $|T(s)|$ and $|T_1(s)|$ have the same degree. Therefore P and P_1 have the same order.

On multiplying out, equation (3.3) gives

$$\begin{pmatrix} T_1 & U_1 \\ -V_1 & W_1 \end{pmatrix} = \begin{pmatrix} MTN & M(TY + U) \\ -(V - XT)N & XTY - VY + XU + W \end{pmatrix} \tag{3.5}$$

Therefore

$$G_1 = V_1T_1^{-1}U_1 + W_1 = (V - XT)NN^{-1}T^{-1}M^{-1}M(TY + U) + XTY - VY + XU + W \tag{3.6}$$

$$= VT^{-1}U + W = G \tag{3.7}$$

This completes the proof.

EXERCISE 3.1 Show that any relation of strict system equivalence can be generated by elementary operations of the following types

(i) Multiply any one of the first r rows (resp. columns) by a non-zero constant.
(ii) Add a multiple, by a polynomial, of any one of the first r rows (resp. columns) to any other row (resp. column).
(iii) Interchange any two among the first r rows (resp. columns).

Interpret these operations in terms of their effect on the system equations. [The unimodular matrices $\begin{pmatrix} M & 0 \\ X & I \end{pmatrix}$ and $\begin{pmatrix} N & Y \\ 0 & I \end{pmatrix}$ can be represented respectively by a product of elementary row and column operations (Chapter 1, Section 1). Note the significance of the zero blocks. This result is important in later developments.]

EXERCISE 3.2 Give an alternative proof of the second part of Theorem 3.1, relying on equation (2.29). [Use the result of Exercise 3.1.]

Theorem 3.2 Any polynomial system matrix $P(s)$ can be brought by strict system equivalence to the form

$$\begin{pmatrix} I_{r-n} & 0 \\ 0 & P_1(s) \end{pmatrix} \tag{3.8}$$

where $P_1(s)$ is a system matrix in state-space form.

Proof Let P be given by equation (3.1), and choose M and N in (3.5) so that T_1 is in Smith form. Because $|T_1|$ has degree n, there will be at least $r - n$ unit entries on the principal diagonal of T_1 (in the leading positions) if $r > n$. Subtract such multiples of the first $r - n$ rows from the last m rows of the system matrix that the entries in the first $r - n$ columns of the system matrix are zero except for unit entries on the principal diagonal. Operate similarly on columns. The system matrix now has the form

$$\left(\begin{array}{cc|c} I_{r-n} & 0 & 0 \\ 0 & T_2 & U_2 \\ \hline 0 & -V_2 & W_2 \end{array}\right) \tag{3.9}$$

where T_2 is the remaining part of T_1. Now choose an M_1 and N_1 in (3.5),

$$M_1 = \begin{pmatrix} I_{r-n} & 0 \\ 0 & M_2 \end{pmatrix}, \quad N_1 = \begin{pmatrix} I_{r-n} & 0 \\ 0 & N_2 \end{pmatrix} \tag{3.10}$$

where M_2 and N_2 are such that $M_2T_2N_2 = sI_n - A$. Finally subtract such multiples of rows $r - n + 1, r - n + 2, \ldots, r$ from the last m rows that these last rows are independent of s, and similarly for the last l columns. The system matrix now has the form

$$\left(\begin{array}{cc|c} I_{r-n} & 0 & 0 \\ 0 & sI_n - A & B \\ \hline 0 & -C & D(s) \end{array}\right) \tag{3.11}$$

Corollary 1 If $P(s)$ gives rise to a proper $G(s)$, then $D(s)$ in (3.11) will be zero.

Corollary 2 In (3.11), A may be chosen in first natural normal form or second natural normal form (Chapter 1, Section 1.1).

Corollary 3 In the preceding part of this section, the matrices M, N, X, Y, T, U, V, W may be restricted to have real coefficients. If M, N, X, Y are allowed to have complex coefficients, A in (3.11) may be chosen in Jordan normal form (Chapter 1, Section 1.1).

Example 3.1 From equations (2.11) there is obtained the system matrix

$$P(s) = \left(\begin{array}{cc|c} s+1 & s^3 & 0 \\ 0 & s+1 & 1 \\ \hline -1 & 0 & 0 \end{array}\right) \tag{3.12}$$

Subtract $(s^2 - s + 1)$ times the second row from the first row to give

$$\left(\begin{array}{cc|c} s+1 & -1 & -s^2 + s - 1 \\ 0 & s+1 & 1 \\ \hline -1 & 0 & 0 \end{array}\right) \tag{3.13}$$

Now add $(s - 2)$ times the first column to the last to give

$$\left(\begin{array}{cc|c} s+1 & -1 & -3 \\ 0 & s+1 & 1 \\ \hline -1 & 0 & -s+2 \end{array}\right) \tag{3.14}$$

These operations will be seen to coincide with those in Example 2.5, but they are here suggested much more naturally by the form of the system matrix. It is also clear from this example that the operations in the proof of Theorem 3.2 will not usually give the simplest way of obtaining the desired form. Usually it is better to work entirely, as here, with elementary operations. Many different routes will lead to the desired goal.

EXERCISE 3.3 Set up the system matrix from equations (2.7) and (2.8) and bring it to state-space form by strict system equivalence.

EXERCISE 3.4 Show that the following are invariant under strict system equivalence:

(i) The Smith form of $P(s)$.
(ii) The Smith form of $T(s)$.
(iii) The Smith form of $\begin{pmatrix} T(s) & U(s) \end{pmatrix}$.
(iv) The Smith form of $\begin{pmatrix} T(s) \\ -V(s) \end{pmatrix}$.

As a consequence of Theorem 3.2,

$$\begin{pmatrix} M & 0 \\ X & I_m \end{pmatrix}\begin{pmatrix} T & U \\ -V & W \end{pmatrix}\begin{pmatrix} N & Y \\ 0 & I_l \end{pmatrix} = \left(\begin{array}{cc|c} I_{r-n} & 0 & 0 \\ 0 & sI_n - A & B \\ \hline 0 & -C & D(s) \end{array}\right) \tag{3.15}$$

and we also have

$$\begin{pmatrix} T & U \\ -V & W \end{pmatrix}\begin{pmatrix} \bar{\xi} \\ -\bar{u} \end{pmatrix} = \begin{pmatrix} 0 \\ -\bar{y} \end{pmatrix} \tag{3.16}$$

and

$$\begin{pmatrix} sI_n - A & B \\ -C & D(s) \end{pmatrix}\begin{pmatrix} \bar{x} \\ -\bar{u} \end{pmatrix} = \begin{pmatrix} 0 \\ -\bar{y} \end{pmatrix} \tag{3.17}$$

from which it is easy to obtain

$$\begin{pmatrix} 0_{r-n,1} \\ \bar{x} \end{pmatrix} = N^{-1}(s)[\bar{\xi} + Y(s)\bar{u}] \tag{3.18}$$

Equation (3.18) shows how the state vector x is related to ξ, u, and their derivatives. Different choices of N and Y will lead to different state vectors.

Example 3.2 In obtaining (3.14) from (3.12) the matrix

$$\begin{pmatrix} N & Y \\ 0 & I_l \end{pmatrix} = \left(\begin{array}{cc|c} 1 & 0 & s-2 \\ 0 & 1 & 0 \\ \hline 0 & 0 & 1 \end{array}\right) \tag{3.19}$$

was used. The state vector is then

$$\begin{pmatrix} x_1 \\ x_2 \end{pmatrix} = \begin{pmatrix} \xi_1 + \dfrac{du}{dt} - 2u \\ \xi_2 \end{pmatrix} \tag{3.20}$$

as in (2.15).

3.2 System similarity

If a system matrix $P(s)$ in state-space form is transformed by strict system equivalence, the result will not generally be in state-space form. If H is a non-singular constant matrix (that is, a matrix not depending on s), then the transformation

$$\begin{pmatrix} H^{-1} & 0 \\ 0 & I_m \end{pmatrix} \begin{pmatrix} sI_n - A & B \\ -C & D(s) \end{pmatrix} \begin{pmatrix} H & 0 \\ 0 & I_l \end{pmatrix} \tag{3.21}$$

evidently preserves the state-space form. This transformation is called *system similarity*. Strict system equivalence clearly includes system similarity as a special case, and it follows that the system order (as is obvious) and the transfer function matrix are both invariant under system similarity. Also equation (3.18) shows that the original state vector x is related to the state vector x_1 after the transformation by

$$x = Hx_1 \tag{3.22}$$

That is, the transformation can be represented as a change of basis in the state space (Chapter 1, Section 4).

Theorem 3.3 Two system matrices $P(s)$ and $P_1(s)$ in state-space form are system similar if and only if they are strictly system equivalent.

Proof Necessity is obvious. To prove sufficiency let P and P_1 be related by

$$\begin{pmatrix} M & 0 \\ X & I_m \end{pmatrix} \begin{pmatrix} sI_n - A & B \\ -C & D \end{pmatrix} \begin{pmatrix} N & Y \\ 0 & I_l \end{pmatrix} = \begin{pmatrix} sI_n - A_1 & B_1 \\ -C_1 & D_1 \end{pmatrix} \tag{3.23}$$

and assume first that A and A_1 are both in first natural normal form. Then from (3.5),

$$M(sI_n - A)N = sI_n - A_1 \tag{3.24}$$

$$M[(sI_n - A)Y + B] = B_1 \tag{3.25}$$

$$[C - X(sI_n - A)]N = C_1 \tag{3.26}$$

From equation (3.24) it follows that A and A_1 have the same invariant polynomials (compare Chapter 1, Theorem 1.7) and as they are both in first natural normal form they are equal. Because P and P_1 give the same transfer function matrix

$$C(sI_n - A)^{-1}B + D(s) = C_1(sI - A)^{-1}B_1 + D_1(s) \tag{3.27}$$

and on equating non-negative powers of s, $D_1(s) = D(s)$.

We may write (Chapter 1, Theorem 1.8)

$$M(s) = (sI_n - A)M_1(s) + M_0 \tag{3.28}$$

$$N(s) = N_1(s)(sI_n - A) + N_0 \tag{3.29}$$

where M_1, N_1 are polynomial matrices and M_0, N_0 are non-singular constant matrices, and then $N_0 = M_0^{-1}$ and (3.24) can be replaced by

$$M_0(sI - A)M_0^{-1} = sI - A \tag{3.30}$$

Write

$$X_a(s) = \{X(s)M^{-1}(s) - CN_1(s)\}M_0 \tag{3.31}$$

$$Y_a(s) = M_0^{-1}\{N^{-1}(s)Y(s) + M_1(s)B\} \tag{3.32}$$

and consider the product

$$\begin{pmatrix} M_0 & 0 \\ X_a(s) & I_m \end{pmatrix}\begin{pmatrix} sI_n - A & B \\ -C & D(s) \end{pmatrix}\begin{pmatrix} M_0^{-1} & Y_a(s) \\ 0 & I_l \end{pmatrix} = \begin{pmatrix} Q_{11} & Q_{12} \\ Q_{21} & Q_{22} \end{pmatrix} \tag{3.33}$$

By direct evaluation using (3.5) and (3.30),

$$Q_{11} = M_0(sI_n - A)M_0^{-1} = sI_n - A \tag{3.34}$$

Also

$$Q_{12} = M_0[(sI_n - A)M_0^{-1}\{N^{-1}(s)Y(s) + M_1(s)B\} + B] \tag{3.35}$$

$$= (sI_n - A)N^{-1}(s)Y(s) + \{(sI_n - A)M_1(s) + M_0\}B \tag{3.36}$$

$$= M(s)(sI_n - A)Y(s) + M(s)B \tag{3.37}$$

$$= B_1 \tag{3.38}$$

where (3.36) uses (3.30), (3.37) uses (3.24) with $A_1 = A$, and (3.38) uses (3.25). In the same way

$$Q_{21} = -C_1 \tag{3.39}$$

and 3.33 becomes

$$\begin{pmatrix} M_0 & 0 \\ X_a(s) & I_m \end{pmatrix}\begin{pmatrix} sI_n - A & B \\ -C & D(s) \end{pmatrix}\begin{pmatrix} M_0^{-1} & Y_a(s) \\ 0 & I_l \end{pmatrix} = \begin{pmatrix} sI_n - A & B_1 \\ -C_1 & Q_{22} \end{pmatrix} \tag{3.40}$$

from which

$$\{X_a(s)(sI_n - A) - C\}M_0^{-1} = -C_1 \tag{3.41}$$

Write

$$X_a(s) = X_0 + X_1 s + \cdots + X_q s^q \tag{3.42}$$

and equate powers of s in (3.41),

$$\left.\begin{aligned} s^{q+1} X_q M_0^{-1} &= 0 \\ s^q (X_{q-1} - X_q A) M_0^{-1} &= 0 \\ \vdots \qquad\qquad & \\ s(X_0 - X_1 A) M_0^{-1} &= 0 \end{aligned}\right\} \tag{3.43}$$

which gives

$$X_q = 0 = X_{q-1} = \cdots = X_0 \tag{3.44}$$

and

$$X_a(s) = 0 \tag{3.45}$$

In the same way

$$Y_a(s) = 0 \tag{3.46}$$

and finally (3.40) gives $Q_{22} = D(s) = D_1(s)$. Therefore the two system matrices, with A and A_1 both in first natural normal form, are system similar, with the transforming matrix $H = M_0^{-1}$.

Now suppose that P and P_1 are given as in the statement of the theorem. By system similarity bring the matrix A in P to first natural normal form. Do the same for P_1. The two resulting matrices are system similar by what was proved above. The theorem follows.

3.3 System equivalence

A third form of system matrix is now introduced. It is written as in (3.1)

$$P(s) = \begin{pmatrix} T(s) & U(s) \\ -V(s) & W(s) \end{pmatrix} \tag{3.47}$$

but now T, U, V, W are matrices over the rational functions. A system matrix of this type is said to be in *rational form* (*third form* elsewhere). The matrices T, U, V, W are $r \times r$, $r \times l$, $m \times r$, $m \times l$, but no order is associated with P, and no restriction is placed on the magnitude of r. As before, it is assumed that $|T(s)| \not\equiv 0$. The class of rational system matrices clearly includes the class of polynomial system matrices, and therefore also includes the class of system matrices in state-space form.

The relation of *system equivalence* is generated by the following operations, which may be carried out in any order and any (finite) number of times:

(i) Multiply any one of the first r rows (resp. columns) by a rational function, not identically zero.
(ii) Add a multiple, by a rational function, of any one of the first r rows (resp. columns) to any other row (resp. column).
(iii) Interchange any two among the first r rows (resp. columns).
(iv) Add a row and column to P to form

$$\begin{pmatrix} 1 & 0 \\ 0 & P \end{pmatrix} \tag{3.48}$$

or, if the system matrix has the form (3.48), delete the first row and first column to leave P.

Comparison with Exercise 3.1 shows that system equivalence includes strict system equivalence (and therefore also system similarity) as a special case.

It is easy to show that the transfer function matrix is invariant under system equivalence: the proof in Theorem 3.1 holds with only minor changes when the matrices are given the appropriate interpretation.

Theorem 3.4 The transfer function matrix $G(s)$ is a standard form for system matrices under system equivalence.

Proof Operate on P as follows:

$$\begin{pmatrix} T & U \\ -V & W \end{pmatrix} \to \begin{pmatrix} I & T^{-1}U \\ -V & W \end{pmatrix} \to \begin{pmatrix} I & T^{-1}U \\ 0 & VT^{-1}U + W \end{pmatrix}$$
$$= \begin{pmatrix} I & T^{-1}U \\ 0 & G \end{pmatrix} \to \begin{pmatrix} I & 0 \\ 0 & G \end{pmatrix} \to G \tag{3.49}$$

Theorem 3.5 Two system matrices are system equivalent if and only if they give rise to the same transfer function matrix.

Proof If the two matrices are system equivalent they give rise to the same $G(s)$, as was noted above. If they give rise to the same $G(s)$ each can be reduced to $G(s)$ by system equivalence, and the desired result follows.

EXERCISE 3.5 Show that equation (2.29) holds also for rational system matrices.

EXERCISE 3.6 Show that system equivalence is an equivalence relation (R & S, p. 8). So also is strict system equivalence (and therefore also system similarity).

[Notice that because operation (iv) of system equivalence is not permitted in strict system equivalence, a separate proof is needed for the latter.]

4 Reduction of order

Suppose that two polynomial system matrices P and P_1 exist, with different orders n and n_1, $n > n_1$, but giving rise to the same transfer function matrix. By Theorem 3.5 there is a transformation of system equivalence which carries P into P_1. By this means a system of lower order is obtained from P which yet gives the same G. In this section a method will be given for generating certain transformations of this type.

Example 4.1 In equation (1.19), put $R_1 = R_2 = 0$. The system matrix becomes

$$P(s) = \left(\begin{array}{cc|c} s & 0 & 1/L_2 \\ 0 & s & 1/L_2 \\ \hline -1 & 0 & 0 \end{array}\right) \tag{4.1}$$

with order 2, and $G(s) = 1/L_2 s$. The system matrix

$$P_1(s) = \left(\begin{array}{c|c} s & 1/L_2 \\ \hline -1 & 0 \end{array}\right) \tag{4.2}$$

has order 1 and gives the same $G(s)$. The following operations of system equivalence transform P into P_1: divide column 2 by s, subtract $1/L_2$ times column 2 from column 3, transpose columns 1 and 2, transpose rows 1 and 2, reduce the size of the matrix by 1. The successive results are

$$\begin{pmatrix} s & 0 & 1/L_2 \\ 0 & 1 & 1/L_2 \\ -1 & 0 & 0 \end{pmatrix} \to \begin{pmatrix} s & 0 & 1/L_2 \\ 0 & 1 & 0 \\ -1 & 0 & 0 \end{pmatrix} \to \begin{pmatrix} 0 & s & 1/L_2 \\ 1 & 0 & 0 \\ 0 & -1 & 0 \end{pmatrix}$$

$$\to \begin{pmatrix} 1 & 0 & 0 \\ 0 & s & 1/L_2 \\ 0 & -1 & 0 \end{pmatrix} \to \begin{pmatrix} s & 1/L_2 \\ -1 & 0 \end{pmatrix} \tag{4.3}$$

Suppose then that a polynomial system matrix $P(s)$ is given as in (3.1), and let the $r \times (r + l)$ polynomial matrix $(T(s)\ U(s))$ have rank less than r for some value s_0 of s. This is the same thing as saying that s_0 is a zero of an invariant polynomial in the Smith form of $(T(s)\ U(s))$. Since rank $(T(s_0)\ U(s_0))$ is less than r, there is a non-singular constant matrix K such that in

$$K(T(s_0) \quad U(s_0)) \tag{4.4}$$

one of the rows is zero, say row i. Assume that T and U (as usual) have real coefficients, and suppose first that s_0 is real. Then K will be real, and in the matrix

$$K(T(s) \quad U(s)) \tag{4.5}$$

row i will be divisible by $s - s_0$. Accordingly, the following operations of system equivalence will transform P into a polynomial system matrix P_1 with real coefficients and with order one less than P:

(i) Multiply P on the left by $\begin{pmatrix} K & 0 \\ 0 & I_m \end{pmatrix}$.

(ii) Divide row i by $s - s_0$.

If s_0 is complex, say $s_0 = \sigma_0 + i\omega_0$, K in (4.4) will be complex. Write

$$M(s) = \operatorname{Re} K - \frac{\sigma_0}{\omega_0} \operatorname{Im} K + \frac{s}{\omega_0} \operatorname{Im} K \tag{4.6}$$

so that $M(s)$ is a polynomial matrix with real coefficients. Notice that K has the form

$$K = \begin{pmatrix} 1 & 0 & \dots & 0 & \dots & 0 \\ 0 & 1 & \dots & 0 & \dots & 0 \\ \vdots & \vdots & & \vdots & & \vdots \\ k_{i1} & k_{i2} & \dots & 1 & \dots & k_{ir} \\ \vdots & \vdots & & \vdots & & \vdots \\ 0 & 0 & \dots & 0 & \dots & 1 \end{pmatrix} \tag{4.7}$$

so that $M(s)$ is a matrix having the same form as K, but with k_{ij} replaced by

$$m_{ij} = \operatorname{Re} k_{ij} - \frac{\sigma_0}{\omega_0} \operatorname{Im} k_{ij} + \frac{s}{\omega_0} \operatorname{Im} k_{ij}, \qquad j = 1, 2, \dots, i-1, i+1, \dots r \tag{4.8}$$

It follows that $M(s)$ has determinant 1.

Now from (4.6),

$$M(s_0) = \operatorname{Re} K - \frac{\sigma_0}{\omega_0} \operatorname{Im} K + \frac{\sigma_0 + i\omega_0}{\omega_0} \operatorname{Im} K = K \tag{4.9}$$

$$M(\bar{s}_0) = \operatorname{Re} K - \frac{\sigma_0}{\omega_0} \operatorname{Im} K + \frac{\sigma_0 - i\omega_0}{\omega_0} \operatorname{Im} K = \bar{K} \tag{4.10}$$

and because

$$\bar{K}(T(\bar{s}_0) \quad U(\bar{s}_0)) = \overline{K(T(s_0) \quad U(s_0))} \tag{4.11}$$

row i in the matrix

$$M(s)\big(T(s) \quad U(s)\big) \tag{4.12}$$

will be divisible by $(s - s_0)(s - \bar{s}_0)$.

The following operations of system equivalence will therefore transform P into a polynomial system matrix P_1 with real coefficients and with order two less than P:

(i) Multiply P on the left by $\begin{pmatrix} M(s) & 0 \\ 0 & I_m \end{pmatrix}$.

(ii) Divide row i by $(s - s_0)(s - \bar{s}_0)$.

If there is a value s_1 of s for which

$$\text{rank}\begin{pmatrix} T(s) \\ -V(s) \end{pmatrix} < r \tag{4.13}$$

the order of P can be reduced by analogous operations on columns. Also it is clear that if T, U or T, V have complex coefficients, and if rank $(T \; U) < r$ or rank $\begin{pmatrix} T \\ -V \end{pmatrix} < r$ for some value s_0 of s, then again a reduction of order can be made in P, but the result in general will have complex coefficients. We thus have proved the following result.

Theorem 4.1 If in the $(r + m) \times (r + l)$ polynomial system matrix $P(s)$ either rank $\big(T(s)\; U(s)\big)$ or rank $\big(T^{\mathrm{T}}(s) \; -V^{\mathrm{T}}(s)\big)$ is less than r for some value s_0 of s, then there exists an $(r + m) \times (r + l)$ polynomial system matrix $P_1(s)$ of lower order than $P(s)$ giving rise to the same transfer function matrix. If P has real coefficients, then P_1 can be chosen with real coefficients.

Corollary 1 There exists a P_1 as in Theorem 4.1 unless both $\big(T(s)\; U(s)\big)$ and $\big(T^{\mathrm{T}}(s) - V^{\mathrm{T}}(s)\big)$ have Smith form $(I_r\; 0)$.

Proof The polynomial matrix $T(s)$ is non-singular, so the two Smith forms have no zero entries on the leading diagonal. If either Smith form has a non-unit entry on the leading diagonal, the corresponding matrix $(T\; U)$ or $(T^{\mathrm{T}} - V^{\mathrm{T}})$ has rank less than r for some s.

Corollary 2 If $T_1(s)$ is the leading $r \times r$ block matrix of P_1, then $|T(s)|/|T_1(s)|$ is a polynomial, and its roots are precisely the roots which were removed from P.

EXERCISE 4.1 Find a system matrix of lower order giving rise to the same transfer function as

$$\left(\begin{array}{cc|c} 1 & s+1 & s+2 \\ s+1 & s+3 & 6 \\ \hline -1 & 0 & 0 \end{array}\right) \tag{4.14}$$

Comment on the stability of the reduced and the original system. [Notice that if $(T(s_0)\ U(s_0))$ or $(T^{\mathrm{T}}(s_0)\ -V^{\mathrm{T}}(s_0))$ has rank less than r, $|T(s_0)| = 0$. Therefore we need only test for linear dependence at the n roots of $|T(s)| = 0$.]

It will be noticed that after reduction of the order, the value of r can always be reduced. This is most easily seen by bringing $T_1(s)$ to Smith form by suitable operations of strict system equivalence on P_1, when there will be $r - n_1 > 0$ unit entries (at least) on the principal diagonal. In practice this step can usually be avoided.

If in P both $(T\ U)$ and $(T^{\mathrm{T}}\ -V^{\mathrm{T}})$ have Smith form $(I_r\ 0)$, the above procedure does not allow any further reduction of the order. The system matrix P (or the system itself) is then said to have *least order*, or to be *irreducible* or *minimal*. At this stage we have no way of knowing whether some other technique would lead to a P_1 having lower order but giving the same G. In fact it will be shown in Chapter 3 that no such P_1 exists. This justifies the term 'least order'.

The terms 'irreducible' and 'minimal' were in use for state-space systems, and do not apply very naturally to more general systems. They refer only to the order, and an irreducible system may still be capable of simplification in other respects. Consequently 'least order' will be preferred in this book.

Example 4.2 The system matrix

$$P(s) = \left(\begin{array}{cc|c} s+1 & s^3+2s^2 & s^2+1 \\ s^2+3s+2 & s^4+4s^3+4s^2+s+2 & s^3+2s^2+s+3 \\ \hline s^2+3s+1 & s^4+4s^3+4s^2-1 & s^3+2s^2+s+2 \end{array}\right) \tag{4.15}$$

has least order. It is strictly system equivalent to

$$P_1(s) = \left(\begin{array}{cc|c} s+1 & 0 & 1 \\ 0 & s+2 & 1 \\ \hline -1 & -1 & 0 \end{array}\right) \tag{4.16}$$

5 Decoupling zeros

The procedure given in Section 4 allows a deeper analysis to be made of the structure of a system. Suppose that in P the matrix $(T\ U)$ has Smith form $S = (Q\ 0)$. Then $|Q(s)| = D_r(s)$, where D_r is the greatest monic common divisor of minors of order r in $(T\ U)$. Let the roots of $|Q(s)| = D_r(s) = 0$ be $\beta_1, \beta_2, \ldots, \beta_b$: for brevity these will be called the zeros of S. Then

$$(T(\beta_i)\quad U(\beta_i)), \quad i = 1, 2, \ldots, b \tag{5.1}$$

has rank less than r. If β_1 is real, a factor $s - \beta_1$ can be removed from P. When this is done, each $r \times r$ minor of $K(T\ U)$ is divided by $s - \beta_1$. The greatest common divisor of all $r \times r$ minors in $K(T\ U)$ is the same as in $(T\ U)$. It follows that removal of $s - \beta_1$ from P divides D_r by $s - \beta_1$. Similarly if β_1 is complex the removal of a complex pair of roots divides D_r by the factors removed.

The removal of any roots leaves D_r still divisible by the remaining roots. Consequently all the roots $\beta_1, \beta_2, \ldots \beta_b$ can be removed in turn, and when this has been done the resulting $(T_1\ U_1)$ will have Smith form $(I_r\ 0)$. The complete set of roots $\{\beta_i\}$ will be called the *input-decoupling zeros* (or more briefly the *i.d. zeros*) of P.

Example 5.1 Let P be the system matrix

$$P(s) = \left(\begin{array}{cc|c} s+1 & 0 & 1 \\ 0 & s+2 & 0 \\ \hline -1 & -1 & 0 \end{array}\right) \tag{5.2}$$

There is one i.d. zero, $\beta = -2$, and the corresponding mode is decoupled from the input. That is to say, let the system start from zero initial conditions at $t = 0$, and let an arbitrary input be applied on the interval $0 < t \leqslant t_0$; thereafter let the input be zero. Then the free motion for $t > t_0$ will be $c_1 e^{-(t-t_0)} + c_2 e^{-2(t-t_0)}$, and c_2 will be zero regardless of the input. This partially explains the term 'input-decoupling zero' which will be more fully justified in Chapter 5. Notice that the system matrix

$$P_1(s) = \left(\begin{array}{cc|c} (s+1) & 0 & 1 \\ -s(s+1) & s+2 & 2 \\ \hline -1 & -1 & -1 \end{array}\right) \tag{5.3}$$

also has one i.d. zero, $\beta = -2$, though this is less obvious than before. The matrix P_1 was obtained from P by strict system equivalence (compare Exercise 3.4).

In a similar way suppose that in P the matrix $(T^{\mathrm{T}}\ -V^{\mathrm{T}})$ has as the zeros of its

Smith form $\gamma_1, \gamma_2, \ldots, \gamma_c$. These will be called the *output-decoupling zeros*, or *o.d. zeros*, of P.

Further, let a system matrix P be given with i.d. zeros $\{\beta_i\}$. Let these all be removed by the procedure of Section 4, and call the resulting system matrix P_1. In P_1, let the matrix $(T_1^T \; -V_1^T)$ have as the zeros of its Smith form $\theta_1, \theta_2, \ldots, \theta_t$. It will be shown below that

$$\{\gamma_i\} \supseteq \{\theta_i\} \tag{5.4}$$

Let the members of the set $\{\gamma_i\} - \{\theta_i\}$ be $\delta_1, \delta_2, \ldots, \delta_d$: these will be called the *input-output-decoupling zeros*, or *i.o.d. zeros*, of P. Clearly $\{\gamma_i\} \supseteq \{\delta_i\}$, and it will be shown below that

$$\{\beta_i\} \supseteq \{\delta_i\} \tag{5.5}$$

Example 5.2 The system matrix

$$P(s) = \left(\begin{array}{cccc|c} I_2 & 0 & 0 & & 0 \\ 0 & s(s+1) & 0 & & s \\ 0 & 0 & s(s+2) & & s \\ \hline 0 & -1 & -1 & & 0 \end{array}\right) \tag{5.6}$$

has $\{\beta_i\} = \{0, 0\}$ and $\{\gamma_i\} = \{0\}$. When the i.d. zeros are removed, the result is

$$P_1(s) = \left(\begin{array}{cc|c} s+1 & 0 & 1 \\ 0 & s+2 & 1 \\ \hline -1 & -1 & 0 \end{array}\right) \tag{5.7}$$

which has no o.d. zero. Hence $\{\theta_i\} = \varnothing$, the null set, and $\{\delta_i\} = \{0\}$. Notice that the two i.d. zeros of P can be associated with the factors s in rows 3 and 4. When either of these is removed the o.d. zero disappears. Therefore in (5.5), which here becomes

$$\{0, 0\} \supseteq \{0\} \tag{5.8}$$

we cannot identify the root $\delta_1 = 0$ with either of the two roots $\beta_1 = 0, \beta_2 = 0$. The members of the sets $\{\beta_i\}$, $\{\gamma_i\}$, etc. must be regarded as complex numbers which are not further individualized.

Example 5.3 The system matrix

$$P(s) = \left(\begin{array}{cccc|c} I_2 & 0 & 0 & & 0 \\ 0 & s(s+1) & 0 & & s \\ 0 & 0 & s(s+2) & & 1 \\ \hline 0 & -1 & -s & & 0 \end{array}\right) \tag{5.9}$$

has $\{\beta_i\} = \{0\}$ and $\{\gamma_i\} = \{0\}$. However, when the i.d. zero is removed the o.d. zero remains, and so $\{\delta_i\} = \varnothing$. This shows that mere coincidence of the numerical values of a β_i and a γ_i does not imply that they coincide with a δ_i.

Examples 5.2 and 5.3 show that there is a difficulty in identifying individually the members of the sets $\{\beta_i\}$, $\{\gamma_i\}$, etc. For this reason we shall avoid speaking of unions and intersections of such sets, which would imply the ability to distinguish the elements separately. The sets, in what follows, are to be regarded as collections of complex numbers.

We now define a set $\{\alpha_i\}$ as follows. Let the zeros of $|T(s)|$ be $\eta_1, \eta_2, \ldots, \eta_n$, and let $\{\beta_i, \gamma_i\}$ be the set having as its members $\beta_1, \beta_2, \ldots, \beta_b, \gamma_1, \gamma_2, \ldots, \gamma_c$. Because $\{\gamma_i\} \supseteq \{\delta_i\}$, it follows that

$$\{\beta_i, \gamma_i\} \supseteq \{\delta_i\} \tag{5.10}$$

Also it is shown below that

$$\{\eta_i\} \supseteq \{\beta_i, \gamma_i\} - \{\delta_i\} \tag{5.11}$$

Define $\alpha_1, \alpha_2, \ldots, \alpha_a$ by

$$\{\alpha_i\} = \{\eta_i\} - \{\{\beta_i, \gamma_i\} - \{\delta_i\}\} \tag{5.12}$$

The members of $\{\alpha_i\}$ will be called the *poles* of $G(s)$, where $G(s)$ is the transfer function matrix corresponding to P. The reason for the name will become clear in Chapter 3.

Example 5.4 The system matrix

$$P(s) = \left(\begin{array}{cccc|c} I_2 & 0 & 0 & & 0 \\ 0 & s^2(s+1) & s(s+2) & & -s \\ 0 & 0 & s+2 & & 1 \\ \hline 0 & 0 & -1 & & 0 \end{array}\right) \tag{5.13}$$

has

$$\{\eta_i\} = \{0, 0, -1, -2\}$$
$$\{\beta_i\} = \{0\}$$
$$\{\gamma_i\} = \{0, 0, -1\}$$
$$\{\theta_i\} = \{0, -1\}$$
$$\{\delta_i\} = \{0\}$$
$$\{\alpha_i\} = \{-2\}$$

and the transfer function is $1/(s+2)$, having one pole at -2.

Theorem 5.1 With the above notation,

$$\{\gamma_i\} \supseteq \{\theta_i\}, \qquad \{\beta_i\} \supseteq \{\delta_i\} \qquad \text{and} \qquad \{\eta_i\} \supseteq \{\beta_i, \gamma_i\} - \{\delta_i\}$$

Proof Consider the transformation of the matrix $\begin{pmatrix} T \\ -V \end{pmatrix}$ as β_1 is removed from $(T\ U)$. The two following operations are involved.

(i) Replacement of T by KT or by $M(s)T$, where $|M(s)| = 1$.

This leaves the Smith form of $\begin{pmatrix} T \\ -V \end{pmatrix}$ unchanged, because for example

$$\begin{pmatrix} MT \\ -V \end{pmatrix} = \begin{pmatrix} M & 0 \\ 0 & I \end{pmatrix} \begin{pmatrix} T \\ -V \end{pmatrix} \tag{5.14}$$

and the first matrix on the right has determinant 1.

(ii) Division of one row of $T_0 = KT$ or $T_0 = MT$ by $s - \beta_1$. This removes a zero from some of the $r \times r$ minors of $\begin{pmatrix} T_0 \\ -V \end{pmatrix}$ and may leave its Smith form unchanged or may remove a zero equal to β_1 from the Smith form.

Successive operations of these types will therefore never add a new member to $\{\gamma_i\}$ and may possibly delete some members. What remains of the set $\{\gamma_i\}$ after all the β_i have been removed from $(T\ U)$ is $\{\theta_i\}$, and it follows that $\{\gamma_i\} \supseteq \{\theta_i\}$.

If a member γ_j of the set $\{\gamma_i\}$ is deleted by the removal of β_k, then as above $\gamma_j = \beta_k$. Since the members of $\{\gamma_i\}$ which are removed in this way are the set $\{\delta_i\}$ it follows that $\{\beta_i\} \supseteq \{\delta_i\}$.

Every β_i is a zero of every $r \times r$ minor of $(T\ U)$, and is therefore a zero of $|T|$: that is, a member of $\{\eta_i\}$. Similarly every $\gamma_i \in \{\eta_i\}$. Whenever a β_i is removed from $(T\ U)$, a zero η_i equal to β_i is removed from $|T|$. After all the β_i have been removed, T has become T_1. Subsequent removal of the θ_i from $\begin{pmatrix} T_1 \\ -V_1 \end{pmatrix}$ removes further zeros from $|T_1|$, each equal to a $\theta_i \in \{\gamma_i\}$. This shows that

$$\{\eta_i\} \supseteq \{\beta_i, \theta_i\} = \{\beta_i, \gamma_i\} - \{\delta_i\} \tag{5.15}$$

After all the θ_i have been removed, the remaining zeros of $|T_1|$ are the set $\{\alpha_i\}$.

Corollary 1 Every i.o.d. zero is both an i.d. zero and an o.d. zero.

Proof In the proof of Theorem 5.1 it was shown that $\{\beta_i\} \supseteq \{\delta_i\}$, and by definition $\{\gamma_i\} \supseteq \{\delta_i\}$.

Corollary 2 $a + b + c = n + d$.

Proof This is merely a consequence of definition (5.12).

Corollary 3 The α_i are the remaining zeros of $|T(s)|$ after P has been reduced as far as possible.

It will be clear from the method of proof that the set $\{\delta_i\}$ could have been obtained by first removing all the $\{\gamma_i\}$ from $\begin{pmatrix} T \\ -V \end{pmatrix}$. In the resulting system matrix P_2, the Smith form of $(T_2\ U_2)$ will possibly have less zeros than are contained in the set $\{\beta_i\}$. The missing members form the set $\{\delta_i\}$.

Theorem 5.2 Under strict system equivalence (and therefore also under system similarity) the following are invariant.

(i) The set $\{\alpha_i\}$ of poles of G, a in number.
(ii) The set $\{\beta_i\}$ of i.d. zeros of P, b in number.
(iii) The set $\{\gamma_i\}$ of o.d. zeros of P, c in number.
(iv) The set $\{\delta_i\}$ of i.o.d. zeros of P, d in number.
(v) The set $\{\eta_i\}$ of zeros of $|T|$, n in number.

Proof If

$$\begin{pmatrix} M_0 & 0 \\ X & I_m \end{pmatrix}\begin{pmatrix} T & U \\ -V & W \end{pmatrix}\begin{pmatrix} N & Y \\ 0 & I_l \end{pmatrix} = \begin{pmatrix} T_1 & U_1 \\ -V_1 & W_1 \end{pmatrix} \tag{5.16}$$

then

$$M_0TN = T_1 \tag{5.17}$$

$$M_0(T\quad U)\begin{pmatrix} N & Y \\ 0 & I_l \end{pmatrix} = (T_1\quad U_1) \tag{5.18}$$

$$\begin{pmatrix} M_0 & 0 \\ X & I \end{pmatrix}\begin{pmatrix} T \\ -V \end{pmatrix}N = \begin{pmatrix} T_1 \\ -V_1 \end{pmatrix} \tag{5.19}$$

From (5.17) follows (v), from (5.18) follows (ii), and from (5.19) follows (iii). Let the result of removing the set of β_i from $(T\ U)$ be represented by the polynomial matrix $Q(s)(T\ U)$, where $Q(s)$ is a rational matrix. Then from (5.18)

$$M_0QM_0^{-1}(T_1\quad U_1) = M_0Q(T\quad U)\begin{pmatrix} N & Y \\ 0 & I_l \end{pmatrix} \tag{5.20}$$

so that $M_0QM_0^{-1}(T_1\ U_1)$ is a polynomial matrix which has the same Smith form as $Q(T\ U)$, that is $(I_r\ 0)$, and so represents the result of one way of removing the β_i from $(T_1\ U_1)$. Also from (5.19)

$$\begin{pmatrix} M_0QM_0^{-1}T_1 \\ -V_1 \end{pmatrix} = \begin{pmatrix} M_0 & 0 \\ XQ^{-1} & I_m \end{pmatrix}\begin{pmatrix} QT \\ -V \end{pmatrix}N \tag{5.21}$$

and Q^{-1} can be generated as a product of operations of the type K^{-1}, $M^{-1}(s)$ (where K and $M(s)$ are as in Section 4) and multiplication of a row by $s - \beta_i$. Consequently Q^{-1} is a polynomial matrix, and (5.21) shows that

$$\begin{pmatrix} M_0QM_0^{-1}T_1 \\ -V_1 \end{pmatrix} \quad \text{and} \quad \begin{pmatrix} QT \\ -V \end{pmatrix}$$

have the same Smith form; this has as its zeros $\{\theta_i\}$. The results (iv) and (i) now follow from the definition of $\{\delta_i\}$ and from (5.12).

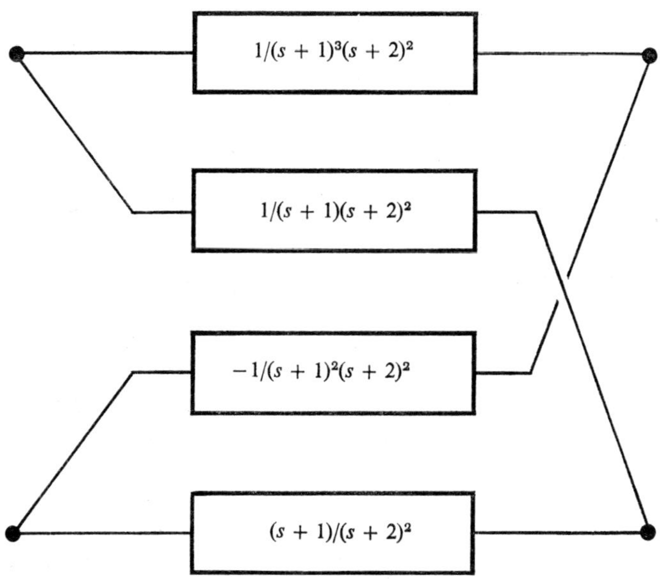

FIG. 3

EXERCISE 5.1 Show that a direct analogue simulation as in Fig. 3 of the transfer function matrix

$$G(s) = \begin{pmatrix} \dfrac{1}{(s+1)^3(s+2)^2} & \dfrac{-1}{(s+1)^2(s+2)^2} \\ \dfrac{1}{(s+1)(s+2)^2} & \dfrac{s+1}{(s+2)^2} \end{pmatrix} \tag{5.22}$$

is represented by the system matrix

$$\left(\begin{array}{ccccc|cc} I_{10} & 0 & 0 & 0 & 0 & 0 & 0 \\ 0 & (s+1)^3(s+2)^2 & 0 & 0 & 0 & 1 & 0 \\ 0 & 0 & (s+1)^2(s+2)^2 & 0 & 0 & 0 & -1 \\ 0 & 0 & 0 & (s+1)(s+2)^2 & 0 & 1 & 0 \\ 0 & 0 & 0 & 0 & (s+2)^2 & 0 & s+1 \\ \hline 0 & -1 & -1 & 0 & 0 & 0 & 0 \\ 0 & 0 & 0 & -1 & -1 & 0 & 0 \end{array}\right) \tag{5.23}$$

and this has

$$\begin{aligned} \{\alpha_i\} &= \{-1, -1, -1, -2, -2, -2\} \\ \{\beta_i\} &= \{-1, -2, -2, -2, -2\} \\ \{\gamma_i\} &= \{-1, -1, -2, -2, -2, -2\} \\ \{\delta_i\} &= \{-2, -2, -2\} \end{aligned} \tag{5.24}$$

so that $a = 6, b = 5, c = 6, d = 3$.

6 Relatively prime polynomial matrices

There is an alternative way of regarding the relation between a system matrix and its reduced form. Suppose that the matrix $(T\ U)$ is reduced, as in the proof of Theorem 5.2, to $(T_2\ U_1) = Q(s)(T\ U)$. Then $Q^{-1}(s)$ is a polynomial matrix, and

$$T = Q^{-1}T_2, \qquad U = Q^{-1}U_1 \tag{6.1}$$

We shall say that T and U have the *common* (*left*) *divisor* Q^{-1}. Suppose that $\begin{pmatrix} T_2 \\ -V \end{pmatrix}$ is further reduced to $\begin{pmatrix} T_1 \\ -V_1 \end{pmatrix} = \begin{pmatrix} T_2 \\ -V \end{pmatrix} R(s)$, where again $R^{-1}(s)$ is a polynomial matrix. This gives

$$T = (Q^{-1}T_1)R^{-1}, \qquad V = V_1R^{-1} \tag{6.2}$$

and T, V will be said to have a *common* (*right*) *divisor* R^{-1}. The relation between the initial system matrix P and the final matrix P_1 can be written

$$\begin{pmatrix} T & U \\ -V & W \end{pmatrix} = \begin{pmatrix} Q^{-1} & 0 \\ 0 & I_m \end{pmatrix}\begin{pmatrix} T_1 & U_1 \\ -V_1 & W \end{pmatrix}\begin{pmatrix} R^{-1} & 0 \\ 0 & I_l \end{pmatrix} \tag{6.3}$$

On forming the transfer function matrix,

$$G = VT^{-1}U + W = V_1R^{-1}(Q^{-1}T_1R^{-1})^{-1}Q^{-1}U_1 + W \tag{6.4}$$

$$= V_1T_1^{-1}U_1 + W \tag{6.5}$$

The common divisors of T, U and T, V cancel in (6.4).

If two matrices T, U have no common left divisor which is not unimodular [that is, is not regular in the ring of polynomial matrices (Chapter 1, Section 1)] then T, U will be called *relatively* (*left*) *prime*. Similarly if T, V have only unimodular common divisors on the right, they are called *relatively* (*right*) *prime*.

Theorem 6.1 The polynomial matrices T, U, respectively $r \times r$, $r \times l$, are relatively (left) prime if and only if any one of the following equivalent conditions is satisfied.

(i) The rank of $(T(s)\ U(s))$ is r for all (complex) s.
(ii) The Smith form $S(s)$ of $(T(s)\ U(s))$ is $(I_r\ 0)$.
(iii) There exist polynomial matrices $V(s)$, $W(s)$, respectively $l \times r$, $l \times l$, such that the Smith form of the matrix

$$\begin{pmatrix} T(s) & U(s) \\ -V(s) & W(s) \end{pmatrix} \tag{6.6}$$

is I_{r+l}. If T, U have real coefficients, then V, W can be chosen with real coefficients.

(iv) There exist relatively (right) prime polynomial matrices $X(s)$, $Y(s)$, respectively $r \times r$, $l \times r$, such that

$$T(s)X(s) + U(s)Y(s) = I_r \tag{6.7}$$

If T and U have real coefficients, then X and Y can be chosen with real coefficients.

Proof If T, U have common left divisor $L(s)$, then

$$(T\quad U) = L(T_1\quad U_1) \tag{6.8}$$

where L, T_1, U_1 are polynomial matrices and $|L(s)|$ has a zero. This implies that for some s, $(T(s)\ U(s))$ has rank less than r. Therefore (i) implies that T, U are relatively prime. Conversely, if (i) is not true, the procedure of Section 4 gives

$$(T_1\quad U_1) = Q(s)(T\quad U) \tag{6.9}$$

where $Q^{-1}(s)$ is a polynomial matrix.

The rank of $(T(s_0)\ U(s_0))$ is the rank of $S(s_0)$. If $S(s)$ has a zero or a polynomial of degree 1 or more on its leading diagonal, its rank cannot be r for all s. Therefore (ii) implies and is implied by (i).

If (ii) is true, there exist unimodular M, (N_{ij}) such that

$$(T \quad U) = M(I_r \quad 0)\begin{pmatrix} N_{11} & N_{12} \\ N_{21} & N_{22} \end{pmatrix} \tag{6.10}$$

$$= (I_r \quad 0)\begin{pmatrix} MN_{11} & MN_{12} \\ N_{21} & N_{22} \end{pmatrix} \tag{6.11}$$

$$= (I_r \quad 0)\begin{pmatrix} T & U \\ N_{21} & N_{22} \end{pmatrix} \tag{6.12}$$

The last matrix has determinant $\alpha \neq 0$ and we may clearly choose N_{21}, N_{22} so that $\alpha = 1$. Putting $N_{21} = -V$, $N_{22} = W$, this shows that (ii) implies (iii). The converse follows directly.

Since the last matrix in (6.12) is unimodular its inverse Q is a polynomial matrix. Writing Q in partitioned form,

$$(T \quad U)\begin{pmatrix} Q_{11} & Q_{12} \\ Q_{21} & Q_{22} \end{pmatrix} = (I_r \quad 0) \tag{6.13}$$

On putting $Q_{11} = X$, $Q_{21} = Y$, (6.13) gives (6.7), which can be written

$$(T \quad U)\begin{pmatrix} X \\ Y \end{pmatrix} = I_r \tag{6.14}$$

This shows that $(T\ U)$ and $\begin{pmatrix} X \\ Y \end{pmatrix}$ each have rank r for all s. Therefore X, Y are relatively (right) prime by a result analogous to (i), and (iii) implies (iv). Also (iv) implies (6.14) which implies (i), completing the proof.

Corollary Completely analogous results can be stated for relative (right) primeness.

Exercise 6.1 Show that T, U, having complex (resp. real) coefficients are relatively (left) prime if and only if T^*, U^* (resp. T^{T}, U^{T}) are relatively (right) prime. Use this to write out the results implied by the corollary.

When T, U take the special forms $T(s) = sI_n - A$, $U(s) = B$, the above conditions can be used, but a number of further conditions become available.

Theorem 6.2 The polynomial matrices $sI_n - A$, B, respectively $n \times n$, $n \times l$, are relatively (left) prime if and only if any one of the following equivalent conditions is satisfied.

(i) There exists no (real or complex) non-zero vector v which simultaneously satisfies for some i,

$$v^*(\lambda_i I_n - A) = 0 \tag{6.15}$$

and

$$v^*B = 0 \tag{6.16}$$

where $\lambda_1, \lambda_2, \ldots, \lambda_n$ are the eigenvalues of A.

(ii) The complex subspace orthogonal to every column of B does not contain a subspace (other than the trivial subspace $\{0\}$) which is invariant under A^*.

(iii) The $n \times ql$ matrix

$$(B \quad AB \quad \ldots \quad A^{q-1}B) \tag{6.17}$$

has rank n, where q is any integer not less than the degree of the minimal polynomial of A. In particular, we may choose $q = n$.

(iv) The following $qn \times [(q-1)n + ql]$ matrix, with q as in (iii), has rank qn.

$$\begin{pmatrix} I_n & 0 & \ldots & 0 & 0 & 0 & \ldots & 0 & 0 & B \\ -A & I_n & \ldots & 0 & 0 & 0 & \ldots & 0 & B & 0 \\ 0 & -A & \ldots & 0 & 0 & 0 & \ldots & B & 0 & 0 \\ \vdots & \vdots & & \vdots & \vdots & \vdots & & \vdots & \vdots & \vdots \\ 0 & 0 & \ldots & I_n & 0 & B & \ldots & 0 & 0 & 0 \\ 0 & 0 & \ldots & -A & B & 0 & \ldots & 0 & 0 & 0 \end{pmatrix} \tag{6.18}$$

(v) Given any polynomial n-vector $\bar{c}(s)$ with elements of degree $q-1$ or less (q as in (iii)), there exist a polynomial n-vector $\bar{x}(s)$ with elements of degree $q-2$ or less, and a polynomial l-vector $\bar{u}(s)$ with elements of degree $q-1$ or less, such that

$$(sI_n - A)\bar{x}(s) - B\bar{u}(s) = \bar{c}(s) \tag{6.19}$$

If A, B and the coefficients of $\bar{c}(s)$ are real, then the coefficients of $\bar{x}(s)$ and $\bar{u}(s)$ can be chosen real.

(vi) There exist an $n \times n$ polynomial matrix $X(s)$ with elements of degree $q-2$ or less (q as in (iii)) and an $l \times n$ polynomial matrix $Y(s)$ with elements of degree $q-1$ or less, such that

$$(sI_n - A)X(s) + BY(s) = I_n \tag{6.20}$$

If A and B are real, $X(s)$ and $Y(s)$ can be chosen with real coefficients.

Proof If $sI_n - A$ and B are relatively prime, the matrix $(sI_n - A \;\; B)$ has rank n for all s, and no v exists satisfying (6.15) and (6.16). Conversely, if no such v exists, $(sI_n - A \;\; B)$ has rank n for $s = \lambda_1, \lambda_2, \ldots, \lambda_n$. For other values of s, $sI_n - A$ has rank n and therefore so does $(sI_n - A \;\; B)$. This shows that $sI_n - A$ and B are relatively prime.

Taking the complex conjugate transpose of (6.15)

$$(\bar{\lambda}_i I_n - A^*)v = 0 \tag{6.21}$$

and if λ_i is an eigenvalue of A then $\bar{\lambda}_i$ is an eigenvalue of A^*. Therefore v is an eigenvector of A^* and spans a subspace which is invariant under A^* (Chapter 1, Section 1). Conversely it will be shown in the following paragraph that every non-trivial subspace invariant under A^* contains an eigenvector of A^*. From this it follows that (ii) implies and is implied by (i).

Let V be a subspace of dimension t which is invariant under A^*. Given any $v \in V$, there exists a k, $1 \leqslant k \leqslant t$, such that $v, A^*v, \ldots, (A^*)^{k-1}v$ are linearly independent, but for some $a_0, a_1, \ldots, a_{k-1}$,

$$a_0 v + a_1 A^* v + \cdots + a_{k-1}(A^*)^{k-1}v + (A^*)^k v = 0 \tag{6.22}$$

Let μ be a solution of the equation

$$a_0 + a_1\mu + \cdots + a_{k-1}\mu^{k-1} + \mu^k = 0 \tag{6.23}$$

Then the $k + 1$ equations

$$\left.\begin{aligned} \mu b_0 \qquad\quad &= a_0 \\ \mu b_1 - b_0 \quad &= a_1 \\ \vdots \qquad\quad & \\ \mu b_{k-1} - b_{k-2} &= a_{k-1} \\ - b_{k-1} &= 1 \end{aligned}\right\} \tag{6.24}$$

determine a unique solution for the k quantities $b_0, b_1, \ldots, b_{k-1}$. For b_{k-1} can be determined from the last equation, b_{k-2} from the last but one, ..., b_0 from the second: then (6.23) is the condition that the first of equations (6.24) should be satisfied. Substituting in (6.22) and rearranging,

$$(\mu I_n - A^*)\big(b_0 v + b_1 A^* v + \cdots + b_{k-1}(A^*)^{k-1}v\big) = 0 \tag{6.25}$$

The vector multiplied by $\mu I_n - A^*$ is non-zero and lies in V; also it is an eigenvector of A^*. This gives the result needed in the preceding paragraph.

Let v be any non-zero real or complex vector. Then either $v^*B \neq 0$ or $v^*B = 0$. If $v^*B = 0$ and (ii) is true, not every one of

$$v, A^*v, \ldots, (A^*)^{q-1}v \tag{6.26}$$

lies in the subspace orthogonal to the columns of B, because A^* has the same minimal polynomial as A and so $(A^*)^q v, (A^*)^{q+1}v, \ldots$, are linearly dependent on the vectors (6.26). In either case

$$v^*(B \quad AB \quad \ldots \quad A^{q-1}B) \neq 0 \tag{6.27}$$

Therefore (ii) implies (iii). Conversely (iii) implies (6.27) for all v. If B has rank n, then certainly (i) is true. If not choose v so that $v^*B = 0$. Then

$$v^*(AB \quad A^2B \quad \ldots \quad A^{q-1}B) \neq 0 \tag{6.28}$$

which means that the vectors $A^*v, (A^*)^2v, \ldots, (A^*)^{q-1}v$ do not remain in the subspace orthogonal to the columns of B, and so (i) is true.

In the matrix (6.18), add A times the first (block) row to the second, then add A times the second (block) row to the third, and so on. The result is a matrix

$$\left(\begin{array}{c|cccc} I_{(q-1)n} & & 0 & & \\ \hline O_{n,(q-1)n} & B & AB & \ldots & A^{q-1}B \end{array}\right) \tag{6.29}$$

which has the same rank as $(B\ AB \ldots A^{q-1}B)$. This shows that (iii) implies and is implied by (iv).

In (6.19), write

$$\bar{x}(s) = x_0 + x_1 s + \cdots + x_{q-2}s^{q-2} \tag{6.30}$$

$$\bar{u}(s) = -u_0 - u_1 s - \cdots - u_{q-1}s^{q-1} \tag{6.31}$$

$$\bar{c}(s) = c_0 + c_1 s + \cdots + c_{q-1}s^{q-1} \tag{6.32}$$

and equate powers of s to obtain

$$\left.\begin{array}{lllllll} x_{q-2} & & +Bu_{q-1} & & & & = c_{q-1} \\ -Ax_{q-2} + x_{q-3} & & & +Bu_{q-2} & & & = c_{q-2} \\ & & \vdots & & & & \vdots \\ & -Ax_1 + x_0 & & & +Bu_1 & & = c_1 \\ & \quad -Ax_0 & & & & +Bu_0 & = c_0 \end{array}\right\} \tag{6.33}$$

If the matrix (6.18) has rank qn, (6.33) has a solution. Unless (6.18) has rank qn, $\bar{c}(s)$ can be chosen so that (6.33) does not have a solution (Chapter 1, Theorem 1.10). Therefore (iv) implies and is implied by (v).

If e_i is the ith column of I_n, put $\bar{c}(s) = e_i$ in (6.19), and let the corresponding solution be $\bar{x}^{(i)}(s)$, $\bar{u}^{(i)}(s)$. If $X(s)$ and $Y(s)$ are the matrices having as their ith columns $\bar{x}^{(i)}(s)$ and $\bar{u}^{(i)}(s)$ respectively, then X and Y satisfy (6.20) and so (v) implies (vi). Conversely, let (6.20) be satisfied; that is, the following equations have a solution for $i = 1, 2, \ldots, n$,

$$\left.\begin{array}{lllllll} x_{q-2}^{(i)} & & +Bu_{q-1}^{(i)} & & & & = 0 \\ -Ax_{q-2}^{(i)} + x_{q-3}^{(i)} & & & +Bu_{q-2}^{(i)} & & & = 0 \\ & & \vdots & & & & \vdots \\ & -Ax_1^{(i)} + x_0^{(i)} & & & +Bu_1^{(i)} & & = 0 \\ & \quad -Ax_0^{(i)} & & & & +Bu_0^{(i)} & = e_i \end{array}\right\} \tag{6.34}$$

Add A times the first of these to the second, then A times the second to the third, and so on, to obtain a final equation

$$A^{q-1}Bu_{q-1}^{(i)} + A^{q-2}Bu_{q-2}^{(i)} + \cdots + Bu_0^{(i)} = e_i \tag{6.35}$$

or

$$(A^{q-1}B \quad A^{q-2}B \quad \ldots \quad B)\begin{pmatrix} u_{q-1}^{(i)} \\ u_{q-2}^{(i)} \\ \vdots \\ u_0^{(i)} \end{pmatrix} = e_i \tag{6.36}$$

This has a solution for $i = 1, 2, \ldots, n$, which implies (Chapter 1, Section 4) that the matrix $(B\ AB\ \ldots\ A^{q-1}B)$ has column rank n. So (vi) implies (iii), and the proof is complete.

Corollary Analogous results can be stated for the relative (right) primeness of $sI_n - A$ and C. In particular $sI_n - A$, C are relatively (right) prime if and only if the complex subspace orthogonal to every column of C^* does not contain a subspace other than $\{0\}$ which is invariant under A.

EXERCISE 6.2 Write out the remaining results implied by the corollary.

EXERCISE 6.3 If A is in Jordan form show that $sI_n - A$, B are relatively (left) prime if and only if the following condition is satisfied. Let $\lambda_1, \lambda_2, \ldots, \lambda_p$ be the distinct eigenvalues of A. Let $b_{i1}, b_{i2}, \ldots, b_{i,k_i}$ be the rows of B corresponding to those rows of $sI_n - A$ having no entry but $s - \lambda_i$. Then for $i = 1, 2, \ldots, p$ the rows $b_{i1}, b_{i2}, \ldots, b_{i,k_i}$ are linearly independent. State the corresponding result for $sI_n - A$, C. [Notice that A, B, C will usually be complex. This is the reason for including this possibility in Theorem 6.2.]

EXERCISE 6.4 Show that if a system in state-space form has least order, then in the Jordan form of A there can be no more than p blocks, $p = \min\{l, m\}$, corresponding to any one eigenvalue. [Bring A to Jordan form by system similarity and use the result of Exercise 6.3.]

Example 6.1 Let P be an $(r + m) \times (r + l)$ polynomial system matrix having least order. Bring P by strict system equivalence to the form

$$P_1(s) = \begin{pmatrix} S(s) & U_1(s) \\ -V_1(s) & W_1(s) \end{pmatrix} \tag{6.37}$$

in which $S(s)$ is the Smith form of the matrix T in P. Suppose that

$$S(s) = \begin{pmatrix} I_{r-p} & 0 & 0 & \ldots & 0 \\ 0 & \psi_1(s) & 0 & \ldots & 0 \\ 0 & 0 & \psi_2(s) & \ldots & 0 \\ \vdots & \vdots & \vdots & & \vdots \\ 0 & 0 & 0 & \ldots & \psi_p(s) \end{pmatrix} \tag{6.38}$$

where the invariant polynomials $\psi_1, \psi_2, \ldots, \psi_p$ have degree greater than zero. If $\psi_1(s_0) = 0$, then by the division properties of the ψ_i every one of $\psi_2(s_0), \psi_3(s_0), \ldots, \psi_p(s_0)$ is also zero. Since P has least order, the matrix $(S(s_0)\; U_1(s_0))$ has rank r, and consideration of the last p rows shows that $p \leqslant l$. Similarly $p \leqslant m$, and so $p \leqslant \min\{l, m\}$. The number p is the greatest number of elementary divisors corresponding to any one root of $|S| = 0$ (Chapter 1, Section 1.1).

Now bring P by strict system equivalence to the form (3.11). The invariant polynomials of A are $\psi_1, \psi_2, \ldots, \psi_p$, and the number p is the greatest number of Jordan blocks (in the Jordan form of A) corresponding to any one eigenvalue. This gives the result of Exercise 6.4. The number p is also the number of companion matrices in the first natural normal form of A. It is also the greatest number of companion matrices, in the second natural normal form of A, corresponding to any one eigenvalue.

EXERCISE 6.5 Show that the system matrix

$$P(s) = \left(\begin{array}{cccc|c} s+1 & -1 & 0 & 1 & \\ 0 & s+1 & 0 & 0 & \\ 0 & 0 & s & 1 & b \\ 0 & 0 & -1 & s+2 & \\ \hline & & -c^{\mathrm{T}} & & d(s) \end{array}\right) \tag{6.39}$$

cannot have least order for any pair of real or complex column vectors b, c.

EXERCISE 6.6 Let the n-vector x satisfy

$$\dot{x} = Ax + Bu; \qquad x(0) = c \tag{6.40}$$

where c can be chosen arbitrarily. Show that if and only if $sI_n - A$ and B are relatively (left) prime, there exists (formally) an l-vector u of impulses and derivatives of impulses such that $x(0+) = 0$. [Use Theorem (6.2)(v).]

7 Reduction of state-space equations

In Section 4 a method was given by which the order of any $(r + m) \times (r + l)$ polynomial system matrix P can be reduced if P has either i.d. or o.d. zeros. When P is in state-space form, this general method can be systematized in the following way. First we define two types of operation on P.

(i) Interchange rows p, q in P to give P_1, where $\alpha \leqslant p \leqslant \beta$, $\alpha \leqslant q \leqslant \beta$, and α, β are specified integers between 1 and n. Follow this immediately by interchanging columns p, q in P_1 to give P_2.

(ii) Add to row p in P a multiple by γ of row q to give P_1, where $p < q \leqslant n$ and γ is a constant. Follow this immediately by subtracting from column q in P_1 a multiple by γ of column p to give P_2.

These two operations are represented by

$$\begin{pmatrix} I_{p-1} & 0 & 0 & 0 \\ 0 & H^{-1} & 0 & 0 \\ 0 & 0 & I_{n-q} & 0 \\ 0 & 0 & 0 & I_m \end{pmatrix} P \begin{pmatrix} I_{p-1} & 0 & 0 & 0 \\ 0 & H & 0 & 0 \\ 0 & 0 & I_{n-q} & 0 \\ 0 & 0 & 0 & I_l \end{pmatrix} = P_1 \tag{7.1}$$

where in operation (i),

$$H^{-1} = \begin{pmatrix} 0 & 0_{1,k} & 1 \\ 0_{k,1} & I_k & 0_{k,1} \\ 1 & 0_{1,k} & 0 \end{pmatrix}, \quad k = q - p - 1 \tag{7.2}$$

and in operation (ii),

$$H^{-1} = \begin{pmatrix} 1 & 0_{1,k} & \gamma \\ 0_{k,1} & I_k & 0_{k,1} \\ 0 & 0_{1,k} & 1 \end{pmatrix}, \quad k = q - p - 1 \tag{7.3}$$

From this it is clear that both (i) and (ii) are operations of system similarity.

Now operate on the matrix

$$P(s) = \begin{pmatrix} sI_n - A & B \\ -C & D(s) \end{pmatrix} \tag{7.4}$$

according to the following algorithm. The objective of the algorithm is to generate as many rows as possible in P (in the order row n, row $n-1, \ldots$) having as their last non-zero entry a constant, and having only zeros above these entries in the same column. Such rows are clearly linearly independent for all s. The index i will distinguish column $n + l - i$, and j will distinguish row $n - j$.

Algorithm 7.1

(1) Set indices $i = 0$, $j = 0$, and go to (2).

(2) If every element in positions $(1, n + l - i)$, $(2, n + l - i), \ldots, (n - j, n + l - i)$ is zero, go to (6). Otherwise go to (3).

(3) Bring a non-zero element to position $(n - j, n + l - i)$ by operation (i) with $\alpha = 1$, $\beta = n - j$. Go to (4).

(4) By operations of type (ii) add such multiples of the element in position

$(n - j, n + l - i)$ to the elements in positions $(1, n + l - i), (2, n + l - i), \ldots, (n - j - 1, n + l - i)$ that these last are reduced to zero. Go to (5).

(5) Increase i by 1 and j by 1. If $j = n$ the process terminates. If $j < n$, go to (2).

(6) Increase i by 1. If $i - j = l$ the process terminates. If $i - j < l$, go to (2).

It is clear that this process must terminate. When it does so, let $j = n - b$. Then, because $n + l - i > n - j$ throughout the procedure, it is easy to see that $P(s)$ will have been reduced to the form

$$P_3(s) = \left(\begin{array}{cc|c} sI_b - A_{11} & 0_{b,n-b} & 0_{b,l} \\ -A_{21} & sI_{n-b} - A_{22} & B_2 \\ \hline -C_1 & -C_2 & D(s) \end{array}\right) \tag{7.5}$$

where $0 \leqslant b \leqslant n$. For the column operations never affect those columns which have been operated on by stage (4) of the algorithm. In P_3, rows $b + 1, b + 2, \ldots, n$ are linearly independent for all s, because in each of these rows the last non-zero element is a constant, and no two rows have their last non-zero element in the same column. These rows remain linearly independent for all s when the first (block) row and first (block) column are removed. On the other hand the first b rows of P_3 are linearly dependent whenever s is an eigenvalue of A_{11}. From this it follows that b is the number of i.d. zeros, and that these zeros are the eigenvalues of A_{11}.

Because P_3 is system similar to P it gives rise to the same transfer function matrix, which is found from (7.5) to be $C_2(sI_{n-b} - A_{22})^{-1}B_2$. It follows that

$$P_4(s) = \begin{pmatrix} sI_{n-b} - A_{22} & B_2 \\ -C_2 & D(s) \end{pmatrix} \tag{7.6}$$

is a system matrix having no i.d. zeros and giving rise to the same transfer function matrix as $P(s)$. By an entirely analogous procedure the o.d. zeros can be removed from $P_4(s)$. It is easy to check that this introduces no new i.d. zeros, so the resulting system matrix has least order.

If this procedure is to be carried out in a digital computer, notice that we need only operate upon

$$\begin{pmatrix} A & B \\ C & 0 \end{pmatrix} \tag{7.7}$$

rather than $P(s)$. The sign changes are trivial, and sI_n and $D(s)$ play no part in the transformation. Notice also that it is unnecessary to have separate algorithms for removing i.d. zeros and o.d. zeros. If a system matrix is transposed its o.d. zeros become i.d. zeros, and one algorithm therefore suffices for both.

EXERCISE 7.1 Justify the statement in the proof that a set of polynomial vectors is linearly independent for all s if (i) each vector has as its last non-zero element a constant; (ii) no two vectors have their last non-zero element in the same position. [Assume the contrary, and consider the last position in which any of the vectors has a non-zero element.]

EXERCISE 7.2 If in (6.39)

$$b = \begin{pmatrix} 4 \\ 3 \\ 2 \\ 1 \end{pmatrix}, \qquad c = \begin{pmatrix} 1 \\ 2 \\ 3 \\ 4 \end{pmatrix} \tag{7.8}$$

show that at the end of stage (4) in Algorithm 7.1 the matrix (7.7) has taken the successive forms, for $i = 0, 1$,

$$\left(\begin{array}{cccc|c} -1 & 1 & -4 & -1 & 0 \\ 0 & -1 & -3 & -3 & 0 \\ 0 & 0 & -2 & -1 & 0 \\ 0 & 0 & 1 & 0 & 1 \\ \hline 1 & 2 & 3 & 20 & 0 \end{array}\right) \tag{7.9}$$

$$\left(\begin{array}{cccc|c} -1 & 1 & 0 & 0 & 0 \\ 0 & -1 & 0 & 0 & 0 \\ 0 & 0 & -2 & -1 & 0 \\ 0 & 0 & 1 & 0 & 1 \\ \hline 1 & 2 & 10 & 20 & 0 \end{array}\right) \tag{7.10}$$

At the next iteration the condition in stage (2) is satisfied and the procedure stops with $i = 3, j = 2$ in stage (6). [Remember that each row operation in (i) or (ii) must be followed immediately by the corresponding column operation.]

7.1 Decomposition of the state space

The algorithm just presented can be used not only for reduction but also, with slight modifications, to generate a decomposition of the state space.

Proceed exactly as before to isolate the i.d. zeros by producing P_3 as in (7.5). Instead of removing the first b rows and columns, retain these in the matrix. Then, as a preliminary to isolating the o.d. zeros, transpose P_3. The essential features of P_3^T (size of blocks, position of zeros) are shown in Fig. 4.

As in the previous algorithm our aim will be to generate an increasing set of vectors, which are linearly independent for all s, until no more can be found. We wish now, however, to do this without losing the two zero blocks in Fig. 4. Examination of the previous algorithm shows that the only stage which can alter these blocks is (3), and then only if the rows interchanged have $p > b$, $q \leqslant b$.

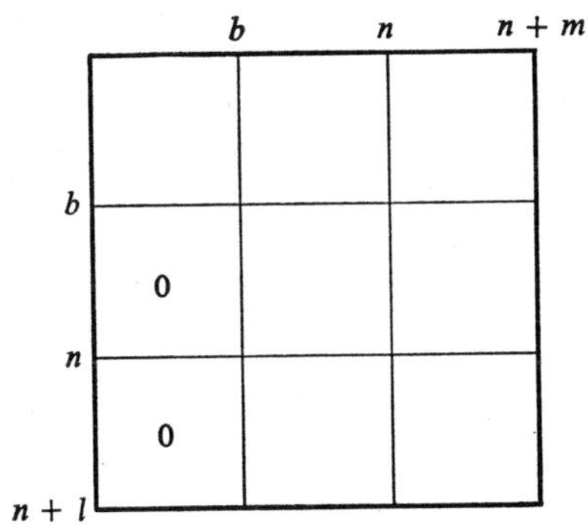

FIG. 4

The algorithm is therefore changed to prevent this possibility. Numbering of the stages is made consecutive with the previous numbering so that the complete set of stages (1) to (18) forms a single algorithm. The index i will distinguish column $n + m - i$, j will distinguish row $n - j$, and k will distinguish row $b - k$.

Algorithm 7.2

(7) Transpose P_3 to form P_3^T as explained above, and go to (8).

(8) Set $i = 0; j = 0$, and go to (9).

(9) If $j = n - b$, set $c - d = 0$ and go to (13). Otherwise go to (10).

(10) If $i - j = m$, set $c - d = n - b - j$ and $i = n + m - b$ and go to (13). Otherwise go to (11).

(11) If $i - j > m$, or if every element in positions $(b + 1, n + m - i)$, $(b + 2, n + m - i), \ldots, (n - j, n + m - i)$ is zero, go to (13). Otherwise go to (12).

(12) Bring a non-zero element to position $(n - j, n + m - i)$ by operation (i) with $\alpha = b + 1, \beta = n - j$. By operations of type (ii) add such multiples of the element in position $(n - j, n + m - i)$ to the elements in positions $(1, n + m - i)$,

$(2, n+m-i), \ldots, (b-k, n+m-i)$, $(b+1, n+m-i)$, $(b+2, n+m-i)$, $\ldots, (n-j-1, n+m-i)$ that these last are reduced to zero. Increase i by 1 and j by 1 and go to (9).

(13) If $k = b$, set $d = 0$ and go to (17). Otherwise go to (14).

(14) If $i - k = n + m - b$, set $d = b - k$ and go to (18). Otherwise go to (15).

(15) If every element in positions $(1, n+m-i)$, $(2, n+m-i), \ldots, (b-k, n+m-i)$ is zero, increase i by 1 and go to (9). Otherwise go to (16).

(16) Bring a non-zero element to position $(b-k, n+m-i)$ by operation (i) with $\alpha = 1$, $\beta = b - k$. By operations of type (ii) add such multiples of the element in position $(b-k, n+m-i)$ to the elements in positions $(1, n+m-i), (2, n+m-i), \ldots, (b-k-1, n+m-i)$ that these last are reduced to zero. Increase i by 1 and j by 1 and go to (9).

(17) If $j = n - b$ or $i - j \geqslant m$, go to (18). Otherwise go to (9).

(18) Transpose the resulting matrix and terminate the process.

Before the final transposition in (18) the matrix will have the form shown in Fig. 5. Rows 1 to b and rows $b + 1$ to n have each been partitioned into two

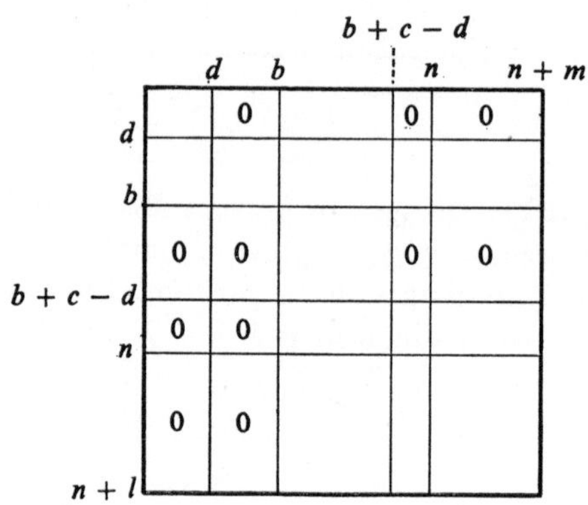

FIG. 5

sets, and the numbers c and d define the sizes of the resulting blocks as shown. The rows $d+1, d+2, \ldots, b, b+c-d+1, b+c-d+2, \ldots, n$, all taken together, are linearly independent for all s because if the second and third (block) columns are interchanged, each of these rows has as its last non-zero element a constant, and no two of the tows terminate at the same column. It is also clear that columns $b, b+1, \ldots, n$ remain linearly independent for all s, because operations (i) and (ii) on these columns, in stages (7) to (18), introduce no other columns and add no multiples of other columns.

On transposing the matrix illustrated in Fig. 5, and allocating symbols to the blocks, we obtain

$$P_5(s) = \left(\begin{array}{ccccc} sI_d - A_{11} & -A_{12} & 0 & 0 & 0 \\ 0 & sI_{b-d} - A_{22} & 0 & 0 & 0 \\ -A_{31} & -A_{32} & sI_{c-d} - A_{33} & -A_{34} & B_3 \\ 0 & -A_{42} & 0 & sI_a - A_{44} & B_4 \\ \hline 0 & -C_2 & 0 & -C_4 & D(s) \end{array}\right) \tag{7.11}$$

where $a = n - b - c + d$. Here b remains equal to the number of i.d. zeros, and these zeros are the eigenvalues of A_{11} and the eigenvalues of A_{22}, taken together. When the i.d. zeros are removed, the number of o.d. zeros falls from c to $c - d$, whence d is the number of i.o.d. zeros, and these zeros are the eigenvalues of A_{11}. Finally a is the number of poles of G, according to the definition in Section 5, and these poles are the eigenvalues of A_{44}. The numbers a, b, c, d introduced in this section are therefore the same as the numbers in Section 5.

As before, Algorithm 7.2 can be applied to the matrix (7.7) rather than (7.4). Though somewhat complicated to describe, the algorithm is easy to code for a digital computer.

Bearing in mind that the transpositions in stages (7) and (18) of Algorithm 7.2 can be omitted if column operations are exchanged appropriately with row operations, it is clear that the matrix (7.11) is system similar to (7.4). On using Theorem 3.2 we may state the following theorem.

Theorem 7.1 Any system matrix in state-space form is system similar to a matrix having the form of P_5 shown in (7.11). Any polynomial system matrix is strictly system equivalent to a matrix having the form

$$\begin{pmatrix} I_{r-n} & 0 \\ 0 & P_5(s) \end{pmatrix} \tag{7.12}$$

In either case, the numbers a, b, c, d in P_5 have the meanings given in Section 5.

If the state vector is partitioned conformably with (7.11), and the differential equations are written out in full, they are:

$$\left.\begin{aligned} \dot{x}_1 &= A_{11}x_1 + A_{12}x_2 \\ \dot{x}_2 &= \qquad\quad A_{22}x_2 \\ \dot{x}_3 &= A_{31}x_1 + A_{32}x_2 + A_{33}x_3 + A_{34}x_4 + B_3u \\ \dot{x}_4 &= \qquad\quad A_{42}x_2 \qquad\qquad + A_{44}x_4 + B_4u \end{aligned}\right\} \tag{7.13}$$

$$y = C_2x_2 + C_4x_4 + D(p)u \tag{7.14}$$

where p is the differential operator. The subspaces spanned by the vectors

$$\xi_1 = \begin{pmatrix} x_1 \\ 0 \\ 0 \\ 0 \end{pmatrix}, \quad \xi_2 = \begin{pmatrix} 0 \\ x_2 \\ 0 \\ 0 \end{pmatrix}, \quad \xi_3 = \begin{pmatrix} 0 \\ 0 \\ x_3 \\ 0 \end{pmatrix}, \quad \xi_4 = \begin{pmatrix} 0 \\ 0 \\ 0 \\ x_4 \end{pmatrix} \tag{7.15}$$

are called respectively:

uncontrollable and unobservable (dimension d);
uncontrollable but observable (dimension $b - d$);
controllable but unobservable (dimension $c - d$);
controllable and observable (dimension $a = n - b - c + d$).

The reasons for these names will become clear in Chapter 5.

Example 7.1 The system matrix (5.23) can be brought to the form (7.11) by strict system equivalence; and when this is done the dimensions of the various subspaces will be

	Controllable	*Uncontrollable*
Observable	6	2
Unobservable	3	3

From equations (7.13) it follows that if $x_2(0) = 0$, then $x_2(t) = 0$, $t > 0$. If $x_1(0) = x_2(0) = 0$, then $x_1(t) = x_2(t) = 0$, $t > 0$. Also each of the vectors ξ_3 and $\xi_1 + \xi_3$ belongs to a subspace invariant under A, for

$$A\xi_1 = \begin{pmatrix} A_{11}x_1 \\ 0 \\ A_{31}x_1 \\ 0 \end{pmatrix}, \quad A\xi_3 = \begin{pmatrix} 0 \\ 0 \\ A_{33}x_3 \\ 0 \end{pmatrix} \tag{7.16}$$

The vectors ξ_1 and ξ_3 are each orthogonal to every column of C^T; compare the corollary of Theorem 6.2.

EXERCISE 7.3 Show that if $u(t) = 0$, $t \geqslant 0$, and $x_2(0) = x_4(0) = 0$, $x_1(0) = c_1$, $x_3(0) = c_3$, then $y(t) = 0$, $t \geqslant 0$. [Use $x(t) = e^{At}c$, expand e^{At} as a series, and use the Cayley–Hamilton theorem to express the series as a matrix polynomial $\phi_0(t)I_n + \phi_1(t)A + \cdots + \phi_{n-1}(t)A^{n-1}$.]

EXERCISE 7.4 Explain the instruction $i = n + m - b$ in stage (10) of the algorithm. [If rows $b + 1, b + 2, \ldots, b + c - d$ are removed in Fig. 5, then so are the corresponding columns. If we continued with $i = n + m - b - c + d$ we should generate linearly independent rows which could fail to remain linearly independent after removing these columns.]

EXERCISE 7.5 Rearrange the blocks in (7.11) in a more symmetrical form. [With A upper or lower triangular.]

8 Number of decoupling zeros

From the particular form of state-space system matrix given in (7.11) it is easy to obtain the following result.

Theorem 8.1 If P is a system matrix in state-space form,

$$P(s) = \begin{pmatrix} sI_n - A & B \\ -C & D(s) \end{pmatrix} \tag{8.1}$$

then the number of i.d. zeros is the rank defect of

$$(B \quad AB \quad \ldots \quad A^{q-1}B) \tag{8.2}$$

and the number of o.d. zeros is the rank defect of

$$\begin{pmatrix} C \\ CA \\ \vdots \\ CA^{q-1} \end{pmatrix} \tag{8.3}$$

In each case, q can be any integer not less than the degree of the minimal polynomial of A, and in particular we may choose $q = n$.

Proof By the transformation

$$P_1 = \begin{pmatrix} H^{-1} & 0 \\ 0 & I_m \end{pmatrix} P \begin{pmatrix} H & 0 \\ 0 & I_l \end{pmatrix} \tag{8.4}$$

bring P to the form of equation (7.5)

$$P_3 = \left(\begin{array}{cc|c} sI_b - A_{11} & 0 & 0 \\ -A_{21} & sI_{n-b} - A_{22} & B_2 \\ \hline -C_1 & -C_2 & D(s) \end{array}\right) \tag{8.5}$$

Then the matrix (8.2) is

$$H\begin{pmatrix} 0 & 0 & \ldots & 0 \\ B_2 & A_{22}B_2 & \ldots & A_{22}^{q-1}B_2 \end{pmatrix} \tag{8.6}$$

and because $sI_{n-b} - A_{22}$ and B_2 are relatively (left) prime, the last $n - b$ rows have rank $n - b$. The rank defect of the matrix (8.6), which is the same as the rank defect of (8.2), is therefore b. The remaining part is proved in a similar way.

EXERCISE 8.1 Show that the number of i.d. zeros of P is (with q as in Theorem 8.1) the rank defect of the $qn \times [(q-1)n + ql]$ matrix

$$\begin{pmatrix} I_n & 0 & \dots & 0 & 0 & 0 & \dots & 0 & 0 & B \\ -A & I_n & \dots & 0 & 0 & 0 & \dots & 0 & B & 0 \\ 0 & -A & \dots & 0 & 0 & 0 & \dots & B & 0 & 0 \\ \vdots & \vdots & & \vdots & \vdots & \vdots & & \vdots & \vdots & \vdots \\ 0 & 0 & \dots & I_n & 0 & B & \dots & 0 & 0 & 0 \\ 0 & 0 & \dots & -A & B & 0 & \dots & 0 & 0 & 0 \end{pmatrix} \tag{8.7}$$

State the corresponding result for o.d. zeros. [Bring the matrix (8.7) to the form (6.29).]

EXERCISE 8.2 If

$$A = \begin{pmatrix} 0 & \dots & 0 & -a_0 \\ 1 & \dots & 0 & -a_1 \\ \vdots & & \vdots & \vdots \\ 0 & \dots & 1 & -a_{n-1} \end{pmatrix}, \qquad B = \begin{pmatrix} b_0 \\ b_1 \\ \vdots \\ b_{n-1} \end{pmatrix} \tag{8.8}$$

verify that

$$(B \quad AB \cdots A^{n-1}B) = b(A) \triangleq b_0 I_n + b_1 A + \cdots + b_{n-1}A^{n-1} \tag{8.9}$$

[Write $b(A)$ in terms of its columns, $b(A) = (v_1, v_2, \dots, v_n)$, and let e_i be the ith column of I_n. Show directly that $v_1 = B$ and then that $v_i = b(A)e_i = b(A)Ae_{i-1}$ $= Ab(A)e_{i-1} = Av_{i-1}$, $i = 2, 3, \dots, n$.]

EXERCISE 8.3 If A, B are as in (8.8) and

$$C = (0 \quad 0 \quad \dots \quad 0 \quad 1) \tag{8.10}$$

show that

$$P(s) = \begin{pmatrix} sI_n - A & B \\ -C & 0 \end{pmatrix} \tag{8.11}$$

has no o.d. zeros and gives

$$G(s) = \frac{b_0 + b_1 s + \cdots + b_{n-1}s^{n-1}}{a_0 + a_1 s + \cdots + a_{n-1}s^{n-1} + s^n} = \frac{b(s)}{a(s)} \tag{8.12}$$

Conclude that the number of common roots of $a(s)$ and $b(s)$ is the rank defect of $b(A)$ defined in (8.9).

EXERCISE 8.4 If the eigenvalues of A in (8.8) are λ_i, $i = 1, 2, \ldots, n$, show that the eigenvalues of $b(A)$ are $b(\lambda_i)$, $i = 1, 2, \ldots, n$. Hence show directly that $a(s)$ and $b(s)$ in (8.12) are relatively prime if and only if $b(A)$ is non-singular. [Bring A to Jordan form.]

Problems

1 Let $G(s)$ arise from

$$P(s) = \begin{pmatrix} sI_n - A & B \\ -C & 0 \end{pmatrix}$$

Show that $s^p G(s) \to 0$ as $s \to \infty$ if and only if $CA^q B = 0$, $q < p$.

2 Give a general procedure for reducing to state-space form the equations

$$Q\dot{x} = Ax + Bu$$

where Q is singular.

3 By strict system equivalence bring $T(s)$ in the polynomial system matrix

$$P(s) = \begin{pmatrix} T(s) & U(s) \\ -V(s) & W(s) \end{pmatrix}$$

to Smith form. Hence show that if the corresponding $G(s)$ is proper it can be written

$$G(s) = \sum_i [v_i(s)u_i^T(s) \bmod \psi_i(s)]/\psi_i(s)$$

where the ψ_i are the invariant polynomials of $T(s)$ and v_i, u_i are certain polynomial column vectors.

4 Show that the $(r + m) \times (r + l)$ rational system matrix

$$P(s) = \begin{pmatrix} T(s) & U(s) \\ -V(s) & W(s) \end{pmatrix}$$

can be reduced by a partial gaussian elimination to the form

$$\begin{pmatrix} P_{11}(s) & P_{12}(s) \\ 0 & P_{22}(s) \end{pmatrix}$$

where P_{11} is $r \times r$ and is upper triangular; and that this can be done without interchanges of columns, and by interchanging only among the first r rows. [Regard P as a matrix over the field of rational functions and note that $|T(s)| \not\equiv 0$.]

5 Show that the transformation induced by the gaussian elimination in Problem 4 is a transformation of system equivalence. Hence show that $P_{22}(s) = G(s)$.

6 Use the result of Problems 4 and 5 to compute the transfer function resulting from $P(s)$ in equation (1.41).

7 Consider $P(s)$ in Problem 4 and the corresponding $G(s)$ as matrices over the rational functions. Show that the rank of $P(s)$ exceeds that of $G(s)$ by r.

8 Let $G(s)$ arise from

$$P(s) = \begin{pmatrix} sI_n - A & B \\ -C & 0 \end{pmatrix}$$

and let $L(s)$ be a polynomial matrix such that $L(s)G(s)$ is proper. Show that $L(s)G(s)$ arises from

$$P_1(s) = \begin{pmatrix} sI_n - A & B \\ -C_1 & 0 \end{pmatrix}$$

where if

$$L(s) = L_0 + L_1 s + \cdots + L_q s^q$$

then

$$C_1 = L_0 C + L_1 CA + \cdots + L_q CA^q$$

9 Let the nonsingular $m \times m$ transfer function matrix $G(s)$ arise from

$$P(s) = \begin{pmatrix} sI_n - A & B \\ -C & 0 \end{pmatrix}$$

and let $G^{-1}(s) = N(s)/d(s)$ where the polynomial $d(s)$ is the least common denominator of the elements of G^{-1}. Show from equation (1.44) that the elements of $N(s)$ have degrees no higher than $n - m + 1$ and $d(s)$ has degree no higher than $n - m$.

10 Show that the $(r + m) \times (r + l)$ polynomial matrix

$$\begin{pmatrix} T^2 & TU \\ -VT & -VU \end{pmatrix}$$

has Smith form

$$\begin{pmatrix} I_r & 0 \\ 0 & 0 \end{pmatrix}$$

if and only if T, U are relatively (left) prime and T, V are relatively (right) prime.

3 Systems of least order

In Chapter 2 it was seen that if a polynomial system matrix has i.d. zeros or o.d. zeros, then its order can be reduced. This is effected by removing the decoupling zeros, and when no more remain the system is said to have least order. This terminology will not be fully justified until Section 3 of this chapter.

The conditions for a system to have least order can be stated in several ways, for example:

(i) there are no decoupling zeros;
(ii) the matrices T, U are relatively (left) prime, and T, V are relatively (right) prime;
(iii) (state-space equations only) the system is observable, and pointwise state controllable.

The first and second of these will be preferred in this book: they are equivalent to a large number of algebraic criteria (Section 6 of Chapter 2). The third is applicable when the system equations are in state-space form, and will become clearer in Chapter 5.

The theory of least-order systems is a specially simple case of the general theory. A more complete account can therefore be given than was possible in Chapter 2. It will be assumed throughout this chapter that $l \leqslant n$, $m \leqslant n$.

1 Standard form under system similarity

Many of the results in this chapter follow from a standard form for polynomial system matrices which will be given in Section 2. As a step towards this we shall first obtain a standard form under system similarity for system matrices in state-space form.

1.1 A standard form

Let an $(n + m) \times (n + l)$ system matrix $P(s)$ be given, in state-space form and having no i.d. zeros,

$$P(s) = \begin{pmatrix} sI_n - A & B \\ -C & D(s) \end{pmatrix} \tag{1.1}$$

Write B in terms of its columns,

$$B = (b_1, b_2, \ldots, b_l) \tag{1.2}$$

Because P has no i.d. zeros, the matrix

$$(B \quad AB \quad \ldots \quad A^{n-1}B) \tag{1.3}$$

has rank n (Chapter 2, Section 8). Consequently the nl vectors

$$b_1, Ab_1, \ldots, A^{n-1}b_1, b_2, Ab_2, \ldots, A^{n-1}b_l \tag{1.4}$$

span E_n. Select a subset of these vectors to make a non-singular matrix

$$H = (b_1, Ab_1, \ldots, A^{p_1-1}b_1, b_2, Ab_2, \ldots, A^{p_l-1}b_l) \tag{1.5}$$

in the following way. Examine the vectors (1.4) in the following order:

$$b_1, b_2, \ldots, b_l, Ab_1, Ab_2, \ldots, A^{n-1}b_l \tag{1.6}$$

Accept each new vector which is examined if it is linearly independent of all those previously accepted; otherwise reject it. When n vectors have been accepted, arrange them in the order (1.5).

Carry out the transformation of system similarity

$$P_1(s) = \begin{pmatrix} H^{-1} & 0 \\ 0 & I_m \end{pmatrix} \begin{pmatrix} sI_n - A & B \\ -C & D(s) \end{pmatrix} \begin{pmatrix} H & 0 \\ 0 & I_l \end{pmatrix} = \begin{pmatrix} sI_n - A_1 & B_1 \\ -C_1 & D(s) \end{pmatrix} \tag{1.7}$$

Write the matrix H in terms of its columns,

$$H = (h_1, h_2, \ldots, h_n) \tag{1.8}$$

Then in AH, columns $h_{p_i+1}, h_{p_i+2}, \ldots, h_{p_{i+1}-1}$ become $h_{p_i+2}, h_{p_i+3}, \ldots, h_{p_{i+1}-1}$, $Ah_{p_{i+1}-1}$, and the last vector can be expressed in the form

$$Ah_{p_{i+1}-1} = \sum_{j=1}^{n} \alpha_j h_j \tag{1.9}$$

because the columns of H span E_n. From this it follows that $A_1 = H^{-1}AH$ has the form

$$A_1 = \begin{pmatrix} Q_{11} & Q_{12} & \cdots & Q_{1l} \\ Q_{21} & Q_{22} & \cdots & Q_{2l} \\ \vdots & \vdots & & \vdots \\ Q_{l1} & Q_{l2} & \cdots & Q_{ll} \end{pmatrix} \tag{1.10}$$

where the Q_{ii} are companion matrices

$$Q_{ii} = \begin{pmatrix} 0 & 0 & \dots & 0 & -q_0^i \\ 1 & 0 & \dots & 0 & -q_1^i \\ 0 & 1 & \dots & 0 & -q_2^i \\ \vdots & \vdots & & \vdots & \vdots \\ 0 & 0 & \dots & 1 & -q_{p_i-1}^i \end{pmatrix}, \quad p_i \times p_i \tag{1.11}$$

and the Q_{ij}, $j \neq i$, have zero entries except perhaps in the last column. If any $p_i = 0$, the corresponding row and column will be absent in (1.10), but they are nevertheless included in the enumeration.

In the matrix $B_1 = H^{-1}B$ the ith column will have one of two forms, according as $p_i > 0$ or $p_i = 0$. If $p_i > 0$, column b_i of B occurs as column $p_1 + p_2 + \dots + p_{i-1} + 1$ of H. Column i of B_1 therefore has all entries zero except for a 1 in row $p_1 + p_2 + \dots + p_{i-1} + 1$. If $p_i = 0$, then column b_i of B is linearly dependent on $b_1, b_2, \dots, b_{i-1}$. Consequently column i of B_1 has zero entries except perhaps in rows $1, p_1 + 1, \dots, p_{i-1} + 1$. The matrix $C_1 = CH$ has no special form, while $D(s)$ is the same in P_1 as in P.

The form described for P_1 is the standard form which was sought. A completely analogous standard form can be obtained for system matrices having no o.d. zeros, in which the roles of rows and of columns are interchanged.

Example 1.1 Let

$$P(s) = \left(\begin{array}{ccc|cc} s-1 & -2 & -3 & 1 & 0 \\ -2 & s-3 & -4 & 0 & 1 \\ -3 & -4 & s-5 & 0 & 0 \\ \hline -1 & 0 & -1 & 0 & 1 \end{array}\right) \tag{1.12}$$

Then

$$(B \quad AB \quad A^2B) = \begin{pmatrix} 1 & 0 & 1 & 2 & 14 & 20 \\ 0 & 1 & 2 & 3 & 20 & 29 \\ 0 & 0 & 3 & 4 & 26 & 38 \end{pmatrix} \tag{1.13}$$

and $p_1 = 2$, $p_2 = 1$,

$$H = \begin{pmatrix} 1 & 1 & 0 \\ 0 & 2 & 1 \\ 0 & 3 & 9 \end{pmatrix} \tag{1.14}$$

$$H^{-1} = \begin{pmatrix} 1 & 0 & -\frac{1}{3} \\ 0 & 0 & \frac{1}{3} \\ 0 & 1 & -\frac{2}{3} \end{pmatrix} \tag{1.15}$$

$$H^{-1}AH = \begin{pmatrix} 0 & \frac{16}{3} & \frac{2}{3} \\ 1 & \frac{26}{3} & \frac{4}{3} \\ 0 & \frac{8}{3} & \frac{1}{3} \end{pmatrix} \tag{1.16}$$

in which there are two companion matrices, respectively 2×2 and 1×1. Also

$$H^{-1}B = \begin{pmatrix} 1 & 0 \\ 0 & 0 \\ 0 & 1 \end{pmatrix} \tag{1.17}$$

$$CH = (1 \quad 4 \quad 0) \tag{1.18}$$

so that the standard form of P is

$$P_1(s) = \left(\begin{array}{ccc|cc} s & -\frac{16}{3} & -\frac{2}{3} & 1 & 0 \\ -1 & s - \frac{26}{3} & -\frac{4}{3} & 0 & 0 \\ 0 & -\frac{8}{3} & s - \frac{1}{3} & 0 & 1 \\ \hline -1 & -4 & 0 & 0 & 1 \end{array}\right) \tag{1.19}$$

EXERCISE 1.1 Check that P and P_1 in Example 1.1 give rise to the same G.

EXERCISE 1.2 Find the matrix H which brings the following system matrix to standard form by system similarity:

$$\left(\begin{array}{cccc|ccc} s & 0 & 0 & 4 & 1 & 1 & 0 \\ -1 & s & 0 & 3 & 0 & 0 & 1 \\ 0 & -1 & s & 2 & 0 & 0 & 0 \\ 0 & 0 & -1 & s+1 & 0 & 0 & 0 \\ \hline 0 & 1 & 0 & 1 & 0 & 0 & 0 \end{array}\right) \tag{1.20}$$

Explain the result.

1.2 An alternative standard form

In Chapter 5 an alternative standard form will be needed. To obtain it, generate H as in Section 1.1, and write H^{-1} in terms of its rows,

$$H^{-1} = \begin{pmatrix} g_1 \\ g_2 \\ \vdots \\ g_n \end{pmatrix} \tag{1.21}$$

Define a new matrix H_1 by

$$H_1 = \begin{pmatrix} g_{p_1} \\ g_{p_1}A \\ \vdots \\ g_{p_1}A^{p_1-1} \\ g_{p_1+p_2} \\ g_{p_1+p_2}A \\ \vdots \\ g_{p_1+p_2+\cdots+p_l}A^{p_l-1} \end{pmatrix} \tag{1.22}$$

As before, if a $p_i = 0$ the corresponding rows will be absent in (1.22), but they are still included in the enumeration.

By the method of construction, $g_{p_1+p_2+\cdots+p_i}$ is orthogonal, when $p_i > 0$, to all the columns of H except $h_{p_1+p_2+\cdots+p_i} = A^{p_i-1}b_i$. Also, at the point where $A^{p_i-1}b_i$ is accepted when forming H, the following vectors have been examined and accepted; or rejected as being linearly dependent on vectors, other than $A^{p_i-1}b_i$, in H:

$$\left.\begin{aligned} A^j b_k, \quad j &= 0, 1, \ldots, p_i - 2 \\ k &= 1, 2, \ldots, l \\ A^{p_i-1}b_1, A^{p_i-1}b_2&, \ldots, A^{p_i-1}b_{i-1} \end{aligned}\right\} \tag{1.23}$$

From this it follows that

$$H_1 B = \begin{pmatrix} B_1' \\ B_2' \\ \vdots \\ B_l' \end{pmatrix} \tag{1.24}$$

where

$$B_i' = \begin{pmatrix} 0_{p_i-1,l} \\ \beta_1, \beta_2, \ldots, \beta_l \end{pmatrix}, \quad \begin{aligned} \beta_1 &= \beta_2 = \cdots = \beta_{i-1} = 0 \\ \beta_i &= 1 \end{aligned} \tag{1.25}$$

When $p_i = 0$ we interpret this last formula to mean that the corresponding B_i' is omitted.

We now show that H_1 defined by (1.22) is non-singular. Suppose to the contrary that there is a v such that

$$v^{\mathrm{T}} H_1 = 0 \tag{1.26}$$

Then

$$v^{\mathrm{T}}H_1b_i = 0, \quad i = 1, 2, \ldots, l \tag{1.27}$$

whence (1.24) gives for the elements v_j of v,

$$v_{p_1} = v_{p_1+p_2} = \cdots = v_n = 0 \tag{1.28}$$

Similarly,

$$v^{\mathrm{T}}H_1Ab_i = 0, \quad i = 1, 2, \ldots, l \tag{1.29}$$

and so

$$v_{p_1-1} = v_{p_1+p_2-1} = \cdots = v_{n-1} = 0 \tag{1.30}$$

where v_{p_1-1} is to be omitted if $p_1 = 1$. In this way it follows that $v = 0$ and so H_1 is non-singular.

It is now easy to see that

$$H_1AH_1^{-1} = \begin{pmatrix} R_{11} & R_{12} & \ldots & R_{1l} \\ R_{21} & R_{22} & \ldots & R_{2l} \\ \vdots & \vdots & & \vdots \\ R_{l1} & R_{l2} & \ldots & R_{ll} \end{pmatrix} \tag{1.31}$$

where the R_{ii} are companion matrices

$$R_{ii} = \begin{pmatrix} 0 & 1 & 0 & \ldots & 0 \\ 0 & 0 & 1 & \ldots & 0 \\ \vdots & \vdots & \vdots & & \vdots \\ 0 & 0 & 0 & \ldots & 1 \\ -r_0^i & -r_1^i & -r_2^i & \ldots & -r_{p_i-1}^i \end{pmatrix} \tag{1.32}$$

and in $R_{ij}, j \neq i$, every row is zero except possibly the last. The alternative standard form for P is

$$P_2(s) = \begin{pmatrix} H_1 & 0 \\ 0 & I_m \end{pmatrix} \begin{pmatrix} sI_n - A & B \\ -C & D(s) \end{pmatrix} \begin{pmatrix} H_1^{-1} & 0 \\ 0 & I_l \end{pmatrix} = \begin{pmatrix} sI_n - A_2 & B_2 \\ -C_2 & D(s) \end{pmatrix} \tag{1.33}$$

in which $B_2 = H_1B$ has the form given in (1.24) and $A_2 = H_1AH_1^{-1}$ has the form given in (1.31). The matrix C_2 has no special form, while $D(s)$ is the same in P_2 as in P.

Example 1.2 In Example 1.1, $p_1 = 2$, $p_2 = 1$. Then from (1.15),

$$H_1 = \begin{pmatrix} g_2 \\ g_2A \\ g_3 \end{pmatrix} = \begin{pmatrix} 0 & 0 & \frac{1}{3} \\ 1 & \frac{4}{3} & \frac{5}{3} \\ 0 & 1 & -\frac{2}{3} \end{pmatrix} \tag{1.34}$$

$$H_1^{-1} = \begin{pmatrix} -\frac{23}{3} & 1 & -\frac{4}{3} \\ 2 & 0 & 1 \\ 3 & 0 & 0 \end{pmatrix} \tag{1.35}$$

$$H_1AH_1^{-1} = \begin{pmatrix} 0 & 1 & 0 \\ \frac{80}{9} & \frac{26}{3} & \frac{10}{9} \\ \frac{8}{3} & 0 & \frac{1}{3} \end{pmatrix} \tag{1.36}$$

which contains two companion matrices, respectively 2×2 and 1×1. Also

$$H_1B = \begin{pmatrix} 0 & 0 \\ 1 & \frac{4}{3} \\ 0 & 1 \end{pmatrix} \tag{1.37}$$

$$CH_1^{-1} = (-\tfrac{14}{3} \quad 1 \quad -\tfrac{4}{3}) \tag{1.38}$$

and so

$$P_2(s) = \left(\begin{array}{ccc:cc} s & -1 & 0 & 0 & 0 \\ -\frac{80}{9} & s - \frac{26}{3} & -\frac{10}{9} & 1 & \frac{4}{3} \\ -\frac{8}{3} & 0 & s - \frac{1}{3} & 0 & 1 \\ \hdashline \frac{14}{3} & -1 & \frac{4}{3} & 0 & 1 \end{array}\right) \tag{1.39}$$

EXERCISE 1.3 Show that P_2 gives rise to the same G as does P given by (1.12).

EXERCISE 1.4 Transform the system matrix given by (1.20) into the standard form P_2 of equation (1.33).

1.3 Minimal indices

The matrix $(sI_n - A \quad B)$ is an example of a *singular pencil of matrices*. The theory of these pencils and their transformations is important for the understanding of system matrices, but it will not be considered here in any depth. We shall, however, show a connection between the numbers p_i introduced in (1.5) and the *minimal indices* of the singular pencil $(sI_n - A \quad B)$.

Minimal indices are defined in the following way. The equation

$$(sI_n - A \quad B)\begin{pmatrix} \bar{x}(s) \\ -\bar{u}(s) \end{pmatrix} = 0 \tag{1.40}$$

admits polynomial solution vectors

$$v(s) = \begin{pmatrix} \bar{x}(s) \\ -\bar{u}(s) \end{pmatrix}$$

Let the highest degree of any element in such a solution be called the degree of the solution.

Example 1.3 The equation

$$\begin{pmatrix} s+1 & 3 & 1 \\ 0 & s+2 & 1 \end{pmatrix}\begin{pmatrix} \bar{x}(s) \\ -\bar{u}(s) \end{pmatrix} = 0 \tag{1.41}$$

admits the solution vector

$$\begin{pmatrix} s-1 \\ s+1 \\ -(s^2+3s+2) \end{pmatrix} \tag{1.42}$$

with degree 2.

Among the solutions of (1.40) there will be at least one, say $v_1(s)$, with lowest degree, say λ_1. A second solution $v_2(s)$ will be linearly dependent on v_1 if there exist polynomials $\alpha(s)$, $\beta(s)$ such that

$$\alpha(s)v_1(s) + \beta(s)v_2(s) = 0 \tag{1.43}$$

If no such α, β exist, then v_1 and v_2 are linearly independent. Among all solutions of (1.40) linearly independent of v_1, there will be at least one, $v_2(s)$, which has lowest degree, say λ_2. Proceeding in this way we may generate successive polynomial vectors $v_1, v_2, v_3, \ldots$, each linearly independent of those which precede it, until no more can be found.

Theorem 1.1 Let $sI_n - A$, B be relatively (left) prime. Then there exist exactly l linearly independent solution vectors $v_i(s)$. The set $\{\lambda_i\}$ of minimal indices is a rearrangement (in order of magnitude) of the set $\{p_i\}$, and $\lambda_1 + \lambda_2 \cdots \lambda_l = n$.

Proof First, the minimal indices are invariant under the transformation

$$K_1(sI_n - A \quad B)K_2 = Q(s) \tag{1.44}$$

where K_1, K_2 are non-singular matrices independent of s, respectively $n \times n$, $(n+l) \times (n+l)$. For if

$$(sI_n - A \quad B)v(s) = 0 \tag{1.45}$$

put

$$K_2^{-1}v(s) = w(s) \tag{1.46}$$

so that $w(s)$ has the same degree as $v(s)$, and then

$$Q(s)w(s) = 0 \tag{1.47}$$

Conversely, if (1.47) is true, then (1.45) is true. Moreover, v_1 and v_2 are linearly dependent if and only if w_1 and w_2 are linearly dependent. Therefore the procedure which generates the minimal indices of $(sI_n - A \; B)$ generates also the minimal indices of $Q(s)$, and the two sets are equal.

Suppose first that no $p_i = 0$. With H_1 as in Section 1.2,

$$H_1(sI_n - A \quad B)\begin{pmatrix} H_1^{-1} & 0 \\ 0 & -I_l \end{pmatrix} = Q(s) = (sI_n - A_2 \quad -B_2) \tag{1.48}$$

where A_2, B_2 have the forms given in equations (1.31), (1.24) respectively. In $Q(s)$, column $n + 1$ has only one non-zero entry, in row p_1. Add such multiplies of column $n + 1$ to each other column that the only non-zero entry in row p_1 is in column $n + 1$. Do the same for rows $p_1 + p_2, \ldots, n$. The result is

$$\begin{pmatrix} sI_{p_1} - R_1 & 0 & \cdots & 0 & -E_1 \\ 0 & sI_{p_2} - R_2 & \cdots & 0 & -E_2 \\ \vdots & \vdots & & \vdots & \vdots \\ 0 & 0 & \cdots & sI_{p_l} - R_l & -E_l \end{pmatrix} \tag{1.49}$$

where

$$R_i = \begin{pmatrix} 0 & 1 & 0 & \cdots & 0 \\ 0 & 0 & 1 & \cdots & 0 \\ \vdots & \vdots & \vdots & & \vdots \\ 0 & 0 & 0 & \cdots & 1 \\ 0 & 0 & 0 & \cdots & 0 \end{pmatrix}, \quad p_i \times p_i \tag{1.50}$$

$$E_i = \begin{pmatrix} 0_{p_i - 1, l} \\ e_1, e_2, \ldots, e_l \end{pmatrix}, \quad \left. \begin{matrix} e_j = 0, & j \neq i \\ e_i = 1 & \end{matrix} \right\} \tag{1.51}$$

Finally rearrange the columns in (1.49) so that column $n + 1$ becomes column

$p_1 + 1$, column $n + 2$ becomes column $p_1 + p_2 + 2$, and so on. The result is a matrix

$$L(s) = \begin{pmatrix} L_1 & 0 & \dots & 0 \\ 0 & L_2 & \dots & 0 \\ \vdots & \vdots & & \vdots \\ 0 & 0 & \dots & L_l \end{pmatrix} \tag{1.52}$$

in which

$$L_i(s) = \begin{pmatrix} s & -1 & 0 & \dots & 0 & 0 \\ 0 & s & -1 & \dots & 0 & 0 \\ \vdots & \vdots & \vdots & & \vdots & \vdots \\ 0 & 0 & 0 & \dots & s & -1 \end{pmatrix}, \quad p_i \times (p_i + 1) \tag{1.53}$$

The equation

$$L_i(s)v(s) = 0 \tag{1.54}$$

has the solution

$$\begin{aligned} v_1(s) &= \alpha_i(s) \\ v_2(s) &= s\alpha_i(s) \\ &\vdots \\ v_{p_i+1}(s) &= s^{p_i}\alpha_i(s) \end{aligned} \tag{1.55}$$

in which $\alpha_i(s)$ can be chosen arbitrarily. It follows that the solution of (1.54) with lowest degree has degree p_i, and all other solutions are linearly dependent upon this.

If $p_j = 0$, the corresponding block L_j is missing from (1.52), and there will be a corresponding column, say column k, of zeros at the right-hand side. This will permit solutions of $L(s)w(s) = 0$ of the form $w_i = 0$, $i \neq k$, $w_k(s) = \beta(s)$. The solution of lowest degree, of this type, has degree 0, and all other solutions of this type are linearly dependent upon it.

From these facts it follows that the minimal indices of $L(s)$ are (in some order) $p_1, p_2, \dots, p_l$. Consequently the same is true for $(sI_n - A \quad B)$. It follows from (1.10) and (1.11) that $p_1 + p_2 + \dots + p_l = n$.

Corollary 1 The minimal indices $\lambda_1, \lambda_2, \dots, \lambda_l$ of $sI_n - A\ B)$, where $sI_n - A$, B are relatively (left) prime, are invariant under system similarity. They are therefore the same for all state-space system matrices of least order generating a given $G(s)$. The minimal indices λ_i may therefore be associated with the input of $G(s)$.

A similar set of minimal indices $\mu_1, \mu_2, \ldots, \mu_m$ arising from $(sI_n - A^T\ C^T)$ may be associated with the output of $G(s)$.

EXERCISE 1.5 Show that if among the minimal indices λ_i of $(sI_n - A\ B)$, where $sI_n - A$, B are relatively (left) prime, the indices $\lambda_1, \lambda_2, \ldots, \lambda_{l-q}$ are zero, but $\lambda_{l-q+1} > 0$, then the rank of B is q. [Show that if $(sI_n - A\ B)\begin{pmatrix} \bar{x} \\ -\bar{u} \end{pmatrix} = 0$, with $\bar{x}$, $\bar{u}$ independent of s, then $\bar{x} = 0$.]

2 Standard form under strict system equivalence

The system matrix P_1 in (1.7) can now be transformed by the following operations of strict system equivalence. Add s times row p_1 to row $p_1 - 1$, then s times row $p_1 - 1$ to row $p_1 - 2, \ldots$, and finally add s times row 2 to row 1. Here p_1 is the dimension of Q_{11} in equation (1.10), and the result of the operations is to eliminate s from the first $p_1 - 1$ columns of P_1 to give P_2, in which the first p_1 columns now have the form

$$\begin{pmatrix}
0 & 0 & \cdots & 0 & s^{p_1} + q_{p_1-1}s^{p_1-1} + \cdots + q_0 \\
-1 & 0 & \cdots & 0 & s^{p_1-1} + q_{p_1-1}s^{p_1-2} + \cdots + q_1 \\
0 & -1 & \cdots & 0 & s^{p_1-2} + q_{p_1-1}s^{p_1-3} + \cdots + q_2 \\
\vdots & \vdots & & \vdots & \vdots \\
0 & 0 & \cdots & -1 & s + q_{p_1-1} \\
0 & 0 & \cdots & 0 & q_{p_1} \\
\vdots & \vdots & & \vdots & \vdots \\
0 & 0 & \cdots & 0 & q_n \\
-c_{11} & -c_{12} & \cdots & -c_{1,p_1-1} & -c_{1,p_1} \\
\vdots & \vdots & & \vdots & \vdots \\
-c_{m1} & -c_{m2} & \cdots & -c_{m,p_1-1} & -c_{m,p_1}
\end{pmatrix} \qquad (2.1)$$

In (2.1), $q_{p_1}, \ldots, q_n$ are elements of $Q_{21}, \ldots, Q_{l1}$, and the c_{ij} are elements of C_1. By adding a suitable multiple (by a polynomial) of column 1 to each other column having a non-zero entry in the second row, all entries in the second row of P_2 except the first can be made zero, and similarly rows $3, 4, \ldots, p_1$ can be made zero except for unit entries in columns $2, 3, \ldots, p_1 - 1$. Now, by adding a suitable multiple of the second row to rows $n + 1, n + 2, \ldots, n + m$, the elements in the first column of P_2 can be made zero except for the unit entry in the second row, and similarly for columns $2, 3, \ldots, p_1 - 1$. Call the result P_3.

These operations leave the first p_1 rows of B_1 unchanged; the only non-zero elements they contain lie in the first row. The first row of P_3 contains also polynomial entries, each corresponding in position to the last column of a companion matrix in P_1. Column p_1 of P_3 contains m polynomial entries generated by the column operations acting on C_1. There are other polynomial entries in the last m rows of P_3, corresponding in position to the last column of each of the remaining companion matrices. The $m \times l$ block at the bottom right of P_3 is still $D(s)$.

A similar procedure can be applied to the remaining companion matrices. Call the result P_4. Then by interchanges among the first n rows and columns of P_4, and by changes of sign, a system matrix P_5 is produced with the following form:

$$P_5 = \left(\begin{array}{cc|c} I_{n-q} & 0_{n-q,q} & 0_{n-q,l} \\ 0_{q,n-q} & T_5(s) & U_5 \\ \hline 0_{m,n-q} & -V_5(s) & D(s) \end{array}\right) \tag{2.2}$$

In P_5, U_5 has q columns each with a single non-zero (in fact unit) entry, where q is the rank of B, no two of which non-zero entries are in the same row: the remaining $l - q$ columns of U_5 are linearly dependent on these q.

If $q < l$, add columns $n - l + 1, n - l + 2, \ldots, n - q$ in P_5 each to one of those among the last l columns in which U_5 has its linearly dependent columns. Then add such multiples of rows $n - l + 1, n - l + 2, \ldots, n - q$ to rows $n - q + 1, n - q + 2, \ldots, n$ that the formerly dependent columns of U_5 are reduced to zero. Reorder rows $n - l + 1, n - l + 2, \ldots, n$ and reallocate the blocks in P_5 to give a matrix

$$P_6 = \left(\begin{array}{cc|c} I_{n-l} & 0_{n-l,l} & 0_{n-l,l} \\ 0_{l,n-l} & T_6(s) & I_l \\ \hline 0_{m,n-l} & -V_6(s) & D(s) \end{array}\right) \tag{2.3}$$

Finally carry out column operations on T_6 and $-V_6$ in such a way that T_6 is brought to lower triangular form T_7, and in each row of T_7 the element of highest degree is on the principal diagonal and is monic: if an element on the principal diagonal is 1, make each other element in the same row zero. This can always be done because $T_6(s)$ is non-singular. Call the result P_7.

The transfer function matrix is obtained in the form

$$G(s) = V_7(s)T_7^{-1}(s) + D(s) \tag{2.4}$$

and because $G - D$ is proper, it follows that

$$\lim_{s \to \infty} V_7(s)T_7^{-1}(s) = 0 \tag{2.5}$$

Bearing in mind Theorem 3.2 of Chapter 2, we may now state the following theorem.

Theorem 2.1 Any $(r + m) \times (r + l)$ polynomial system matrix $P(s)$ having no i.d. zeros is strictly system equivalent to a system matrix of the form

$$\left(\begin{array}{cc|c} I_{r-l} & 0_{r-l,l} & 0_{r-l,l} \\ 0_{l,r-l} & T(s) & I_l \\ \hline 0_{m,r-l} & -V(s) & D(s) \end{array}\right) \tag{2.6}$$

in which

(i) $T(s)$ is lower triangular, and has each element on the principal diagonal monic and of higher degree than any other non-zero element in its own row. If an element on the principal diagonal is 1, each other element in its own row is zero.
(ii) $V(s)T^{-1}(s)$ tends to zero as $s \to \infty$.

Corollary 1 If $P(s)$ has no o.d. zeros it is strictly system equivalent to a system matrix of the form

$$\left(\begin{array}{cc|c} I_{r-m} & 0_{r-m,m} & 0_{r-m,l} \\ 0_{m,r-m} & T_1(s) & U(s) \\ \hline 0_{m,r-m} & -I_m & D(s) \end{array}\right) \tag{2.7}$$

in which

(i) $T_1(s)$ is lower triangular, and has each element on the principal diagonal monic and of higher degree than any other non-zero element in its own column. If an element on the principal diagonal is 1, each other element in its own column is zero.
(ii) $T_1^{-1}(s)U(s) \to 0$ as $s \to \infty$.

Proof This follows in a completely analogous way on interchanging the roles of rows and columns.

Corollary 2 In Theorem 2.1 and Corollary 1, $T(s)$ and $T_1(s)$ could have been chosen upper triangular.

Corollary 3 Any rational transfer function matrix $G(s)$ can be written in the alternative forms

$$G(s) = V(s)T^{-1}(s) + D(s) \tag{2.8}$$

$$G(s) = T_1^{-1}(s)U(s) + D(s) \tag{2.9}$$

where T, T_1, U and V have the properties stated in Theorem 2.1 and Corollary 1, and T, V are relatively (right) prime, T_1, U are relatively (left) prime.

Proof Apply Theorem 2.1 and Corollary 1 to a system matrix of least order giving rise to $G(s)$.

Example 2.1 On applying the above procedure to the matrix in (1.19) the successive stages are:

$$\left(\begin{array}{ccc|cc} 0 & s^2 - \frac{26}{3}s - \frac{16}{3} & -\frac{4}{3}s - \frac{2}{3} & 1 & 0 \\ -1 & s - \frac{26}{3} & -\frac{4}{3} & 0 & 0 \\ 0 & -\frac{8}{3} & s - \frac{1}{3} & 0 & 1 \\ \hline -1 & -4 & 0 & 0 & 1 \end{array}\right) \tag{2.10}$$

$$\left(\begin{array}{ccc|cc} 0 & s^2 - \frac{26}{3}s - \frac{16}{3} & -\frac{4}{3}s - \frac{2}{3} & 1 & 0 \\ -1 & 0 & 0 & 0 & 0 \\ 0 & -\frac{8}{3} & s - \frac{1}{3} & 0 & 1 \\ \hline 0 & -s + \frac{14}{3} & \frac{4}{3} & 0 & 1 \end{array}\right) \tag{2.11}$$

$$\left(\begin{array}{ccc|cc} 1 & 0 & 0 & 0 & 0 \\ 0 & s^2 - \frac{26}{3}s - \frac{16}{3} & -\frac{4}{3}s - \frac{2}{3} & 1 & 0 \\ 0 & -\frac{8}{3} & s - \frac{1}{3} & 0 & 1 \\ \hline 0 & -s + \frac{14}{3} & \frac{4}{3} & 0 & 1 \end{array}\right) \tag{2.12}$$

$$\left(\begin{array}{ccc|cc} 1 & 0 & 0 & 0 & 0 \\ 0 & s^2 - \frac{26}{3}s - \frac{16}{3} & \frac{3}{8}s^3 - \frac{27}{8}s^2 - \frac{9}{4}s & 1 & 0 \\ 0 & -\frac{8}{3} & 0 & 0 & 1 \\ \hline 0 & -s + \frac{14}{3} & -\frac{3}{8}s^2 + \frac{15}{8}s + \frac{3}{4} & 0 & 1 \end{array}\right) \tag{2.13}$$

$$\left(\begin{array}{ccc|cc} 1 & 0 & 0 & 0 & 0 \\ 0 & s^3 - 9s^2 - 6s & -\frac{3}{8}s^2 + \frac{13}{4}s + 2 & 1 & 0 \\ 0 & 0 & 1 & 0 & 1 \\ \hline 0 & -s^2 + 5s + 2 & \frac{3}{8}s - \frac{7}{4} & 0 & 1 \end{array}\right) \tag{2.14}$$

Here upper triangular form was chosen for T, as this gave somewhat less work.

EXERCISE 2.1 Bring the matrix in (1.20) to the standard form (2.6) by strict system equivalence.

EXERCISE 2.2 Bring the matrix in (1.19) to the standard form (2.7) by strict system equivalence.

EXERCISE 2.3 Bring the following system matrix to the form (2.6) by strict system equivalence,

$$P(s) = \left(\begin{array}{cc|c} 1 & 0 & 0 \\ 0 & (s+2)^2 & s+1 \\ \hline 0 & -1 & 0 \end{array}\right) \tag{2.15}$$

EXERCISE 2.4 Show that a rational $m \times l$ transfer function matrix $G(s)$ can be written in the form $G(s) = V(s)T^{-1}(s) + D(s)$ in which $VT^{-1} \to 0$ as $s \to \infty$; D is a polynomial matrix; and either

(i) T has the element on the principal diagonal in each row of higher degree than any other element in the same row, and the degrees of the elements on the principal diagonal (not necessarily in order) are the minimal indices λ_i associated with the input; or

(ii) T has the element on the principal diagonal in each column of no less degree than any other element in the same column of T and of higher degree than any element in the corresponding column of V; and the degrees of the elements on the principal diagonal of T (not necessarily in order) are the minimal indices λ_i associated with the input.

[For (i), omit to make T_6 triangular, but interchange columns. For (ii) use the standard form of Section 1.2 and note that U can finally be made equal to I_l by appropriate operations.]

EXERCISE 2.5 State results for $G(S) = T_1^{-1}(s)U(s) + D(s)$ corresponding to those given in Exercise 2.4 for $G(s) = V(s)T^{-1}(s) + D(s)$. [Apply the results of Exercise 2.4 to $G^{T}(s)$. Note that the minimal indices $\mu_1, \mu_2, \ldots, \mu_m$ associated with the output replace $\lambda_1, \lambda_2, \ldots, \lambda_l$.]

2.1 Uniqueness of the standard form

The standard forms in (2.6) and (2.7) required respectively that P should have no i.d. zeros, or no o.d. zeros. If P has least order it has neither i.d. nor o.d. zeros, and it will be shown that each of the standard forms is then unique.

Theorem 2.2 If the system matrices (2.6) and (2.7) have no decoupling zeros they are each uniquely determined by the transfer function matrix $G(s)$ associated with them. So, too, are the corresponding matrices with T and T_1 upper triangular.

Proof The proof will be given for (2.7) with T_1 lower triangular. Proofs in the other cases follow similar lines. For convenience the subscript on T_1 will be dropped. Let G_q, U_q, D_q be the matrices formed from the first q rows of G, U, D respectively, $1 \leqslant q \leqslant m$. Let T_q be the leading $q \times q$ submatrix of T, and let $M_q = (T_q \; U_q)$. Because T is lower triangular,

$$G_q - D_q = T_q^{-1} U_q \tag{2.16}$$

and this transfer function arises from

$$P_q = \begin{pmatrix} I_{r-m} & 0 & 0 \\ 0 & T_q & U_q \\ 0 & -I_q & 0 \end{pmatrix} \tag{2.17}$$

This system matrix has no i.d. zero, for if it had then P would also have an i.d. zero. Because $G - D$ tends to zero as $s \to \infty$, D is uniquely determined by G and so therefore is D_q, $q = 1, 2, .., m$.

Let $p \geqslant 0$ indices be chosen subject to

$$1 \leqslant i_1 < i_2 < \cdots < i_p \leqslant q \tag{2.18}$$

$$1 \leqslant j_1 < j_2 < \cdots < j_p \leqslant l \tag{2.19}$$

and consider the minor formed from rows $i_1, i_2, \ldots, i_p$ and columns $j_1, j_2, \ldots j_p$ of $G_q - D_q$. By equation (2.29) of Chapter 2 this is

$$\frac{P_q\binom{i_1, i_2, \ldots, i_p}{j_1, j_2, \ldots, j_p}}{|T_q|} \tag{2.20}$$

Because of the special form of P_q, the numerator minor in (2.20) is equal (within a factor ± 1) to the determinant formed from T_q by deleting columns $i_1, i_2, \ldots, i_p$, and then adjoining columns $j_1, j_2, \ldots, j_p$ from U_q. This determinant is a $q \times q$ minor of M_q, and by suitable choice of the indices (2.18) and (2.19) every such $q \times q$ minor except $|T_q|$ can be obtained. Call these $q \times q$ minors Q_k.

Now suppose that there is a second system matrix P' with the form (2.7), having least order, and giving rise to the same G. Distinguish the submatrices of P' by a prime: T', U'_q etc. From what has been said,

$$\frac{Q_k}{|T_q|} = \frac{Q'_k}{|T'_q|}, \quad \text{all } k \tag{2.21}$$

where k indexes all the corresponding $q \times q$ minors formed from M_q and M'_q. From (2.21),

$$|T'_q| Q_k = |T_q| Q'_k, \quad \text{all } k \tag{2.22}$$

Suppose that $|T'_q|$ has a factor not contained in $|T_q|$. Then this factor divides every Q'_k. But this is impossible because if $|T'_q|$ and all the Q'_q vanish simultaneously, P'_q has an i.d. zero. Therefore every factor of $|T'_q|$ is a factor of $|T_q|$, and conversely. Since T_q and T_q' are triangular with monic polynomials on the principal diagonal, this implies that $|T_q| = |T_q'|$, whence $t_{ii} = t_{ii}'$, $i = 1, 2, \ldots, m$. Then (2.21) shows that $Q_k = Q_k'$, all k. Therefore we conclude that every $q \times q$ minor, without exception, formed from M_q is equal to the corresponding minor of M_q'.

As the basis for an induction, assume $M_{q-1} = M_{q-1}'$, and write m_q, m_q' for the qth rows of M_q, M_q' respectively. Each $q \times q$ minor of M_q and of M_q' can be expanded by its last row, and the coefficients in the expansion (which are minors of M_{q-1} or M_{q-1}') are the same in each case. Therefore all $q \times q$ minors formed from the matrix

$$\begin{pmatrix} M_{q-1}(s) \\ m_q(s) - m_q'(s) \end{pmatrix} \tag{2.23}$$

are identically zero. Regarding the polynomial matrix (2.23) as a matrix over the field of rational functions, it follows that its rows are linearly dependent. But the rows of $M_{q-1}(s)$ are linearly independent for all s, because P_{q-1} has no i.d. zeros. Consequently we may write

$$m_q(s) - m_q'(s) = w(s)M_{q-1}(s) \tag{2.24}$$

where $w(s)$ is a row vector of rational functions.

Let α be the highest order among the poles of the elements $w_i(s)$, and let a pole of order α occur at $s = \beta$. If $\alpha > 0$, then in the equation

$$(s - \beta)^\alpha[m_q(s) - m_q'(s)] = [(s - \beta)^\alpha w(s)]M_{q-1}(s) \tag{2.25}$$

the left-hand side is zero at $s = \beta$. The vector $(s - \beta)^\alpha w(s)$ has all its elements finite at $s = \beta$, and has $\gamma \geqslant 1$ of these elements non-zero. Consequently the γ rows of $M_{q-1}(\beta)$, corresponding to the non-zero elements, are linearly dependent. But this is impossible, for β would then be an i.d. zero of P. Consequently the vector $w(s)$ has polynomial elements.

Now consider the ith element in the vector equation (2.24), $1 \leqslant i \leqslant q - 1$,

$$t_{qi} - t_{qi}' = w_i t_{ii} + w_{i+1}t_{i+1,i} + \cdots + w_{q-1}t_{q-1,i} \tag{2.26}$$

Putting $i = q - 1$ in (2.26) we obtain

$$t_{q,q-1} - t_{q,q-1}' = w_{q-1}t_{q-1,q-1} \tag{2.27}$$

and because the degree of $t_{q-1,q-1}$ exceeds that of $t_{q,q-1}$ and that of $t_{q,q-1}'$, it follows that $w_{q-1} = 0$. Then for $i = q - 2$ in (2.26),

$$t_{q,q-2} - t_{q,q-2}' = w_{q-2}t_{q-2,q-2} \tag{2.28}$$

and again $w_{q-2} = 0$. In this way we can show that $w(s) \equiv 0$, and so $m_q = m_q'$.

It follows that if $M_{q-1} = M_{q-1}'$, then $M_q = M_q'$. But certainly $M_1 = M_1'$, for the 1×1 minors of M_1 have been proved equal to the corresponding 1×1 minors of M_1'. By induction $M_m = M_m'$, that is

$$\begin{pmatrix} T(s) & U(s) \end{pmatrix} = \begin{pmatrix} T'(s) & U'(s) \end{pmatrix} \tag{2.29}$$

from which $P(s) = P'(s)$. This concludes the proof.

EXERCISE 2.5 It has been assumed throughout this chapter that $n \geqslant m$, $n \geqslant l$. Show how the results of this section can be extended to the cases $m > n$ or $l > n$ by suitably expanding P. For example show that

$$P = \left(\begin{array}{c|ccc} s+1 & 3 & 2 & 1 \\ \hline -1 & 0 & 0 & 0 \end{array}\right) \tag{2.30}$$

can be expanded to make $r = 3$, then brought to the form

$$\left(\begin{array}{ccc|ccc} 1 & 0 & 0 & 1 & 0 & 0 \\ 0 & 1 & 0 & 0 & 1 & 0 \\ -3 & -2 & s+1 & 0 & 0 & 1 \\ \hline 0 & 0 & -1 & 0 & 0 & 0 \end{array}\right) \tag{2.31}$$

[This generalization does not seem to be needed in applications.]

3 Least order

The results in Section 2 lead immediately to the following basic theorem.

Theorem 3.1 Let P, P_1 be two $(r + m) \times (r + l)$ polynomial system matrices having no decoupling zero. Then P, P_1 are strictly system equivalent if and only if they give the same G.

Proof If P, P_1 are strictly system equivalent, they give the same G by Theorem 3.1 of Chapter 2. If P, P_1 give the same G, bring each by strict system equivalence to the standard form (2.6). The two standard forms arising in this way are identical, being uniquely determined by G. Hence the result. Note that if P_1 has $r_1 \neq r$, the smaller matrix can be expanded trivially to bring it to the same size as the other.

Corollary Let P, P_1 be two system matrices in state-space form having no decoupling zero. Then P, P_1 give the same G if and only if they are system similar.

Proof By Theorem 3.3 of Chapter 2, P, P_1 are system similar if and only if they are strictly system equivalent.

Example 3.1 The two system matrices

$$\left(\begin{array}{cc|c} 1 & 0 & 0 \\ 0 & (s+2)^2 & -1 \\ \hline 0 & s+1 & 0 \end{array}\right) \qquad \left(\begin{array}{cc|c} 1 & 0 & 0 \\ 0 & (s+2)^2 & s+1 \\ \hline 0 & -1 & 0 \end{array}\right) \tag{3.1}$$

have least order and both give rise to the same transfer function matrix

$$G(s) = \frac{s+1}{(s+\ )^2} \tag{3.2}$$

The two matrices are therefore strictly system equivalent, and in fact

$$\left(\begin{array}{cc|c} s+3 & 1 & 0 \\ -(s+2)^2 & -(s+1) & 0 \\ \hline 1 & 0 & 1 \end{array}\right) \left(\begin{array}{cc|c} 1 & 0 & 0 \\ 0 & (s+2)^2 & -1 \\ \hline 0 & s+1 & 0 \end{array}\right)$$

$$\times \left(\begin{array}{cc|c} -(s+1) & -(s+2)^2 & -(s+1) \\ 1 & s+3 & 1 \\ \hline 0 & 0 & 1 \end{array}\right)$$

$$= \left(\begin{array}{cc|c} 1 & 0 & 0 \\ 0 & (s+2)^2 & s+1 \\ \hline 0 & -1 & 0 \end{array}\right) \tag{3.3}$$

Notice how the zero of the transfer function can be transferred by strict system equivalence from V to U. Contrast this with the behaviour of i.d. zeros or o.d. zeros, which remain associated with the input or output respectively under strict system equivalence.

EXERCISE 3.1 Find a transformation of strict system equivalence relating the two matrices

$$\left(\begin{array}{cc|c} s+1 & 2 & 0 \\ 1 & s+2 & s+2 \\ \hline -1 & -1 & 0 \end{array}\right), \qquad \left(\begin{array}{cc|c} 1 & 0 & 0 \\ 0 & s^2+3s & 2s+2 \\ \hline 0 & 1 & 1 \end{array}\right) \tag{3.4}$$

EXERCISE 3.2 Show, without finding an explicit transformation, that the following system matrices are system similar.

$$\left(\begin{array}{ccc|c} s+1 & 2 & 3 & 1 \\ 2 & s+3 & 4 & 0 \\ 3 & 4 & s+5 & 0 \\ \hline -1 & 0 & 0 & s \end{array}\right), \quad \left(\begin{array}{ccc|c} s & 0 & 0 & -1 \\ -1 & s & -6 & 8 \\ 0 & -1 & s+9 & 1 \\ \hline 0 & 0 & -1 & s \end{array}\right) \tag{3.5}$$

The reason for requiring that $r \geqslant n$ in any polynomial system matrix (Section 2.3 of Chapter 2) can now be explained. It will be noticed that in establishing the standard forms (2.6) and (2.7) it was necessary to use the standard form of Section 1.1. Therefore if $P(s)$ is given, and we wish to bring it to the form (2.6) or (2.7), our derivation would require us first to put $P(s)$ into the form of equation (3.8) in Chapter 2. If this is to be done by strict system equivalence, P must have $r \geqslant n$. It might possibly be true that the necessity to go first to state-space form could be avoided, so allowing r to be less than n. The following example shows that in general this is not allowed.

Example 3.2 If we try to find a transformation of strict system equivalence relating

$$\left(\begin{array}{c|c} (s+2)^2 & -1 \\ \hline s+1 & 0 \end{array}\right), \quad \left(\begin{array}{c|c} (s+2)^2 & s+1 \\ \hline -1 & 0 \end{array}\right) \tag{3.6}$$

[compare this with (3.3)] we have

$$\left(\begin{array}{c|c} m & 0 \\ \hline x(s) & 1 \end{array}\right)\left(\begin{array}{c|c} (s+2)^2 & -1 \\ \hline s+1 & 0 \end{array}\right)\left(\begin{array}{c|c} n & y(s) \\ \hline 0 & 1 \end{array}\right) = \left(\begin{array}{c|c} (s+2)^2 & s+1 \\ \hline -1 & 0 \end{array}\right) \tag{3.7}$$

where m, n are non-zero constants, and $x(s)$, $y(s)$ are polynomials. But then

$$x(s)(s+2)^2 n + (s+1)n = -1 \tag{3.8}$$

so that $x(s) = 0$, $n = 0$, and (3.7) cannot be satisfied.

It is Theorem 3.1 which leads to a justification of the term 'least order'. Up to this point, all we have shown is that the procedure of Chapter 2, Section 4, fails to reduce the order of a system matrix having no decoupling zeros. We now have the following much stronger result.

Theorem 3.2 Let $P(s)$ be a polynomial system matrix with order n, having no decoupling zeros and giving rise to a transfer function matrix $G(s)$. Then there exists no polynomial system matrix $P_1(s)$, with order $n_1 < n$, giving rise to $G(s)$.

Proof Suppose, contrary to what is to be proved, that there exists such a matrix P_1. If P_1 has any decoupling zeros, remove them by the procedure of Chapter 2, Section 4: let the result of this procedure (or otherwise the original matrix P_1) be a system matrix P_2 with order $n_2 \leqslant n_1 < n$. Then P_2 has no decoupling zeros and gives rise to G, and (subject to a trivial expansion) we may assume that P, P_2 have the same size. By Theorem 3.1, P_2 is strictly system equivalent to P, and so $n_2 = n$. This contradiction establishes the theorem.

EXERCISE 3.3 Show that to simulate a proper rational transfer function matrix $G(s)$ on an analogue computer, at least n integrators are needed, where n is the least order associated with $G(s)$. [Consider either the state-space equations of the analogue system, or the number of independent initial conditions needed.]

EXERCISE. 3.4 Show that the impedance $1/(s+1)(s+2)$ cannot be obtained from an electrical circuit containing no capacitors, one inductor, and an arbitrary number of resistors.

4 McMillan form of *G*

The deeper structure of a transfer function matrix G can be investigated by means of its *McMillan standard form*, which is defined in the following way. Let $d(s)$ be the monic least common denominator of the elements of G, and write $G = N(s)/d(s)$. Then $N(s)$ can be brought to Smith form by the transformation

$$L(s)N(s)R(s) = S(s) \tag{4.1}$$

where L and R are unimodular polynomial matrices. In the rational matrix $S(s)/d(s)$, there may be common factors between numerator and denominator in the elements on the leading diagonal. Cancel these common factors, and write the result

$$M(s) = [\operatorname{diag}\left(\varepsilon_i(s)/\psi_i(s)\right) \quad 0_{m,l-m}], \quad l > m \tag{4.2}$$

or

$$M(s) = \operatorname{diag}\left(\varepsilon_i(s)/\psi_i(s)\right), \quad l = m \tag{4.3}$$

or

$$M(s) = \begin{pmatrix} \operatorname{diag}\left(\varepsilon_i(s)/\psi_i(s)\right) \\ 0_{m-l,l} \end{pmatrix}, \quad l < m \tag{4.4}$$

Then $M(s)$ is the McMillan form of G.

Notice that $\psi_1(s) = d(s)$, for the contrary would imply that every element of $N(s)$ was divisible by a factor of $d(s)$. There would then be a monic least common denominator for the elements of G with lower degree than $d(s)$, contrary to assumption: compare Corollary 2 of Theorem 4.2 below.

Example 4.1 The matrix

$$G(s) = \begin{pmatrix} \dfrac{1}{s+1} & \dfrac{s+2}{(s+3)^2} \end{pmatrix} \tag{4.5}$$

gives

$$d(s) = (s+1)(s+3)^2 \tag{4.6}$$

$$N(s) = \begin{pmatrix} (s+3)^2 & (s+1)(s+2) \end{pmatrix} \tag{4.7}$$

$$S(s) = \begin{pmatrix} 1 & 0 \end{pmatrix} \tag{4.8}$$

and so

$$M(s) = \begin{pmatrix} \dfrac{1}{(s+1)(s+3)^2} & 0 \end{pmatrix} \tag{4.9}$$

Example 4.2 From

$$G(s) = \begin{pmatrix} \dfrac{1}{(s+1)^2} & \dfrac{1}{(s+1)(s+2)} \\ \dfrac{1}{(s+1)(s+2)} & \dfrac{s+3}{(s+2)^2} \end{pmatrix} \tag{4.10}$$

we obtain

$$d(s) = (s+1)^2(s+2)^2 \tag{4.11}$$

$$N(s) = \begin{pmatrix} (s+2)^2 & (s+1)(s+2) \\ (s+1)(s+2) & (s+1)^2(s+3) \end{pmatrix} \tag{4.12}$$

$$S(s) = \begin{pmatrix} 1 & 0 \\ 0 & (s+1)^2(s+2)^3 \end{pmatrix} \tag{4.13}$$

$$M(s) = \begin{pmatrix} \dfrac{1}{(s+1)^2(s+2)^2} & 0 \\ 0 & s+2 \end{pmatrix} \tag{4.14}$$

Notice that although G is proper, M is not.

Since each invariant polynomial in the Smith form $S(s)$ divides each non-zero polynomial which succeeds it, it follows that if a common factor is cancelled when forming ε_i, then the same factor is cancelled when forming each succeeding ε_j. From this, ψ_i divides $\psi_1, \psi_2, \ldots, \psi_{i-1}$, $i = 2, 3, \ldots, q$, where ε_q is the last non-zero numerator. The zero numerators can be included in the evaluation if desired, with the corresponding denominators put equal to 1

Conversely, no factor is cancelled when forming ε_i which was present but not cancelled in forming any preceding ε_j. Therefore the division property of the invariant polynomials in S is retained, that is ε_i divides $\varepsilon_{i+1}, \varepsilon_{i+2}, \ldots, \varepsilon_q$, $i = 1, 2, \ldots, q - 1$.

Example 4.3 In (4.14), $s + 2$ is regarded as $(s + 2)/1$. Then $\varepsilon_1 = 1$ divides $\varepsilon_2 = s + 2$, and $\psi_2 = 1$ divides $\psi_1 = (s + 1)^2(s + 2)^2$.

Theorem 4.1 Let

$$P(s) = \begin{pmatrix} T(s) & U(s) \\ -V(s) & W(s) \end{pmatrix} \tag{4.15}$$

be an $(r + m) \times (r + l)$ polynomial system matrix of least order giving rise to G. Let the McMillan form of G be M. Let the Smith form of P be S_p, and the Smith form of T be S_T. Then

$$S_p = \begin{pmatrix} I_r & 0_{r,m} \\ 0_{m,r} & \operatorname{diag}(\varepsilon_1, \varepsilon_2, \ldots, \varepsilon_m) \end{pmatrix}, \quad l = m \tag{4.16}$$

with the appropriate changes when $l \neq m$. Also

$$S_T = \begin{pmatrix} I_{r-m} & 0_{r-m,m} \\ 0_{m,r-m} & \operatorname{diag}(\psi_m, \psi_{m-1}, \ldots, \psi_1) \end{pmatrix}, \quad l = m \tag{4.17}$$

where the polynomials ψ_i are arranged in reverse order, and appropriate changes must be made if $l \neq m$.

Proof Assume for simplicity that $l = m$: the other cases are dealt with similarly. By equation (4.1), the system matrix

$$\left(\begin{array}{cc|c} I_{r-m} & 0 & 0 \\ 0 & d(s)I_m & S(s)R^{-1}(s) \\ \hline 0 & -L^{-1}(s) & 0 \end{array}\right) \tag{4.18}$$

gives rise to $G(s)$. On dividing out any common factors in rows $r - m + 1, r - m + 2, \ldots, r$, (4.18) becomes

$$\left(\begin{array}{cc|c} I_{r-m} & 0 & 0 \\ 0 & \operatorname{diag}(\psi_1, \psi_2, \ldots, \psi_m) & \operatorname{diag}(\varepsilon_1, \varepsilon_2, \ldots, \varepsilon_m)R^{-1}(s) \\ \hline 0 & -L^{-1}(s) & 0 \end{array}\right) \tag{4.19}$$

This has least order, because for $i = 1, 2, \ldots, m$, ε_i and ψ_i do not vanish simultaneously, while L^{-1}, R^{-1} are non-singular for all s. Therefore by Theorem 3.1, $P(s)$ can be brought by strict system equivalence to the form (4.19).

Without changing the Smith form of the matrix (4.19) it may be premultiplied by

$$\begin{pmatrix} I_r & 0 \\ 0 & -L(s) \end{pmatrix} \tag{4.20}$$

and postmultiplied by

$$\begin{pmatrix} I_r & 0 \\ 0 & R(s) \end{pmatrix} \tag{4.21}$$

to give

$$\left(\begin{array}{cc:c} I_{r-m} & 0 & 0 \\ 0 & \operatorname{diag}(\psi_1, \psi_2, \ldots, \psi_m) & \operatorname{diag}(\varepsilon_1, \varepsilon_2, \ldots, \varepsilon_m) \\ \hdashline 0 & I_m & 0 \end{array}\right) \tag{4.22}$$

By adding appropriate multiples of the last m rows to rows $r - m + 1, r - m + 2, r$, and then interchanging rows, the matrix (4.22) can be further brought to the form

$$\begin{pmatrix} I_{r-m} & 0 & 0 \\ 0 & I_m & 0 \\ 0 & 0 & \operatorname{diag}(\varepsilon_1, \varepsilon_2, \ldots, \varepsilon_m) \end{pmatrix} \tag{4.23}$$

This is clearly in Smith form, and proves (4.16).

Returning to (4.19), it follows from the divisibility properties of the ψ_i that

$$\begin{pmatrix} I_{r-m} & 0 \\ 0 & \operatorname{diag}(\psi_m, \psi_{m-1}, \ldots, \psi_1) \end{pmatrix} \tag{4.24}$$

is in Smith form. This, since (4.19) is strictly system equivalent to P, is therefore the Smith form of T, which proves (4.17).

Example 4.4 The system matrix

$$P(s) = \left(\begin{array}{ccc:cc} I_2 & 0 & 0 & 0 & 0 \\ 0 & (s+1)^2(s+2) & 0 & s+2 & s+1 \\ 0 & -(s+1) & s+2 & 0 & 1 \\ \hdashline 0 & -1 & 0 & 0 & 0 \\ 0 & 0 & -1 & 0 & 0 \end{array}\right) \tag{4.25}$$

has least order and gives rise to the $G(s)$ in (4.10). The McMillan form of this G has

$$\left.\begin{array}{ll}\varepsilon_1 = 1, & \varepsilon_2 = s + 2\\ \psi_1 = (s+1)^2(s+2)^2, & \psi_2 = 1\end{array}\right\} \tag{4.26}$$

The Smith form of T, from (4.25), is

$$\begin{pmatrix} I_2 & 0 & 0 \\ 0 & 1 & 0 \\ 0 & 0 & (s+1)^2(s+2)^2 \end{pmatrix} \tag{4.27}$$

The Smith form of the system matrix (4.25) is

$$\begin{pmatrix} I_4 & 0 & 0 \\ 0 & 1 & 0 \\ 0 & 0 & s+2 \end{pmatrix} \tag{4.28}$$

Theorem 4.1 explains the name 'poles of G' which was applied in Section 5 of Chapter 2 to those zeros of $|T|$ which remained after all decoupling zeros had been removed. These are now revealed as the zeros of the denominator polynomials $\psi_i(s)$ in the McMillan form of G. In the same way the set of zeros $\{\zeta_i\}$ of the $\varepsilon_i(s)$ will be called the *zeros of G*.

Example 4.5 The transfer function matrix in (4.10) has the set of poles $\{\alpha_i\} = \{-1, -1, -2, -2\}$, and zeros $\{\zeta_i\} = \{-2\}$. Notice that cancellation may occur in $|G|$.

EXERCISE 4.1 If $D(s)$ is a polynomial matrix, show that $G(s) + D(s)$ has the same denominator polynomials ψ_i in its McMillan form as does $G(s)$; but that the numerator polynomials are generally different. [Consider the effect, on the Smith forms of P and T, of adding D.]

Theorem 4.2 Let P be an $(r + m) \times (r + l)$ polynomial system matrix having least order and giving rise to G with McMillan form M as in equations (4.2) to (4.4). Then the monic greatest common divisor of all the $(r + k) \times (r + k)$ minors $P^{i_1, i_2, \ldots, i_k}_{j_1, j_2, \ldots, j_k}$, when these are not all zero, is

$$\varepsilon_1\varepsilon_2 \ldots \varepsilon_k\psi_{k+1}\psi_{k+2} \ldots \psi_q$$

where $k = 1, 2, \ldots, q$ and $q = \min\{m, l\}$.

Proof Assume for simplicity that $l = m$: the other cases are treated similarly. The system matrix P can be brought by strict system equivalence to the form

(4.19). The operations involved in this transformation are those listed in Chapter 2, Exercise 3.1. Of these the first and third clearly leave the divisors of any minor $P^{i_1, i_2, \ldots, i_k}_{j_1, j_2, \ldots, j_k})$ unchanged. The second operation adds, to one row (column) of such a minor, a multiple of some other row (column), which leaves the value of the minor unchanged.

The matrix (4.19) can then be brought to the form (4.22) by the following operations:

(i) Multiply one of rows (columns) $r + 1, r + 2, \ldots, r + m$ by a non-zero constant.
(ii) Interchange two of rows (columns) $r + 1, r + 2, \ldots, r + m$.
(iii) Add to one of rows (columns) $r + 1, r + 2, \ldots, r + m$ a multiple by a polynomial of another of these rows (columns).

Operation (i) leaves unchanged the divisors of each minor $P^{i_1, i_2, \ldots, i_k}_{j_1, j_2, \ldots, j_k})$. Operation (ii) merely interchanges these minors. Operation (iii) leaves each such minor unchanged if both rows (columns) concerned belong to it or do not belong to it. When a row (column) in such a minor has added to it a multiple of a row (column) not belonging to it, its final value is the sum of its original value and a multiple (by a polynomial) of another minor of the same type. Consequently any divisor of all the minors $P^{i_1, i_2, \ldots, i_k}_{j_1, j_2, \ldots, j_k})$ remains a divisor after operation (iii). Since the inverse of this operation is another operation of the same type, it follows that it leaves unchanged the monic greatest common divisor of the minors concerned.

Now in (4.22),

$$P^{i_1, i_2, \ldots, i_k}_{j_1, j_2, \ldots, j_k}) = \pm\, \varepsilon_{j_1}\varepsilon_{j_2} \ldots \varepsilon_{j_k}\psi_{i'_1}\psi_{i'_2} \ldots \psi_{i'_{m-k}} \tag{4.29}$$

where the indices $i'_1, i'_2, \ldots, i'_{m-k}$ are those remaining from $1, 2, \ldots, m$ when $i_1, i_2, \ldots, i_k$ are deleted. Because ε_i divides ε_{i+1}, and ψ_i divides ψ_{i-1}, it follows that all of the minors such as (4.29) are divisible by $\varepsilon_1\varepsilon_2 \ldots \varepsilon_k\psi_{k+1}\psi_{k+2} \ldots \psi_m$. This completes the proof.

Corollary 1 The monic least common denominator of all $k \times k$ minors $G^{i_1, i_2, \ldots, i_k}_{j_1, j_2, \ldots, j_k}$ is the denominator obtained from

$$\frac{\varepsilon_1\varepsilon_2 \ldots \varepsilon_k}{\psi_1\psi_2 \ldots \psi_k} \tag{4.30}$$

when all common factors have been cancelled.

Proof The result follows from equation (2.29) of Chapter 2 on noticing that $P) = \psi_1\psi_2 \ldots \psi_m$.

Corollary 2 If $d(s)$ is the monic least common denominator of all the elements of G, then $d(s) = \psi_1(s)$.

Proof In (4.30), ε_1 and ψ_1 are relatively prime.

Corollary 3 If P is a state-space system matrix of least order giving rise to G, then the minimal polynomial of the plant matrix A is the monic least common denominator $d(s)$ of the elements of G.

Proof The minimal polynomial of A is the invariant polynomial of highest degree of the polynomial matrix $T(s) = sI_n - A$.

Corollary 4 The monic least common denominator of the elements of $G^{-1}(s)$, if this exists, is ε_m.

Proof This follows from equation (1.44) of Chapter 2.

Example 4.6 In (4.25) the 5×5 minors $P^{i}_{j})$ are

$$P^{1}_{1}) = (s+2)^2, \qquad P^{1}_{2}) = (s+1)(s+2), \qquad P^{2}_{1}) = -(s+1)(s+2), \qquad P^{2}_{2}) = (s+1)^2(s+3) \tag{4.31}$$

with greatest common divisor 1. From (4.26), $\varepsilon_1 = 1$, $\psi_2 = 1$, as required by Theorem 4.2. The least common denominator of the elements of G is $(s+1)^2(s+2)^2$, which is ψ_1. The inverse matrix is

$$G^{-1} = \begin{pmatrix} \dfrac{(s+1)^2(s+3)}{s+2} & -(s+1) \\ -(s+1) & s+2 \end{pmatrix} \tag{4.32}$$

having least common denominator $s + 2$, which is ε_2 as in Corollary 4.

Exercise 4.2 Verify the results of Theorem 4.2 and its corollaries for the $G(s)$ in Example 4.1.

4.1 Standard forms for *G*

A standard form for G was given in Corollary 3 of Theorem 2.1. In the light of the above discussion, that result can be restated in a sharper form.

Theorem 4.3 A rational $m \times l$ transfer function matrix $G(s)$ can be written in the standard form

$$G(s) = V(s)T^{-1}(s) + D(s) \tag{4.33}$$

where V, T, and D are uniquely determined by G, and

(i) $T(s)$ and $V(s)$ are relatively (right) prime.
(ii) $|T|$ has degree equal to the least order among systems giving rise to G.

(iii) T is lower triangular, and has each element on the principal diagonal monic and of higher degree than any other non-zero element in its own row. If an element on the principal diagonal is 1, each other element in the same row is zero.
(iv) $V(s)T^{-1}(s) \rightarrow 0$ as $s \rightarrow \infty$, and $D(s)$ is a polynomial matrix.
(v) The Smith form of $T(s)$ is diag $(I_{l-q}, \psi_q, \psi_{q-1}, \ldots, \psi_1)$, where ψ_i are the denominator polynomials in the McMillan form of G and $q = \min\{l, m\}$.
(vi) The Smith form of $V(s) + D(s)T(s)$ is diag $(\varepsilon_1, \varepsilon_2, \ldots, \varepsilon_m)$ when $l = m$, where ε_i are the numerator polynomials in the McMillan form of G, and appropriate changes must be made when $l \neq m$.

Proof Start from a system matrix of least order giving rise to G, and bring it to the standard form (2.6). Then (i), (ii), (iii), and (iv) follow at once. Theorem 4.1 gives (v) and (vi) after a minor transformation. Uniqueness of the representation follows from Theorem 2.2.

Corollary 1 Under the conditions of the theorem, $G(s)$ can be written alternatively

$$G(s) = T_1^{-1}(s)U(s) + D(s) \tag{4.34}$$

where T_1, U, and D are uniquely determined by G and obey conditions analogous to those in the theorem.

Corollary 2 T in (4.33) and T_1 in (4.34) could be chosen upper triangular instead of lower triangular.

5 Realization theory

In Chapter 2, the situation considered was one in which a physical system was given, and its equations could be written down from physical and chemical principles. The transfer function matrix $G(s)$ was then obtained by elementary calculations. The converse problem can occur: given G, find a set of differential equations from which G can arise. Usually the equations must have some special form or properties, so that they can represent a physical system of given type.

This, for example, is the problem of simulating G on an analogue computer, because routine methods are available for setting up the analogue circuit from differential equations in state-space form. The problem of finding an RLC network to realize a given G is another, more difficult, problem of similar type. The greater difficulty arises from the increased constraints which the equations must satisfy if they are to arise from such a network. Because of these constraints it is known, for example, that certain impedances which can be obtained from RLC networks, cannot be obtained from networks (without ideal transformers) having least order.

Often the problem of realization is equated to that of finding state-space equations of least order giving rise to G. As stated, the real problem may be much more difficult, but the difficulties belong in the technology concerned, and not in general systems theory. Attention is therefore restricted here to the simpler problem.

5.1 Computation of $\nu(G)$

It has been seen in Section 3 that associated with any rational transfer function matrix G is a unique least order: this will be denoted by $\nu(G)$. One question which we should like to answer, within the framework of this section, is how to find $\nu(G)$ without going to the length of finding a least-order system giving rise to G. Three solutions to this problem will be given, in the form of algorithms.

Algorithm 5.1 Let a rational $m \times l$ transfer function matrix G be given. Compute the minors of orders $1, 2, \ldots, q$ of G, where $q = \min\{l, m\}$. Find the monic least common denominator $\phi(s)$ of all these minors. The degree of the polynomial $\phi(s)$ is $\nu(G)$.

Proof Let P be a polynomial system matrix of least order giving rise to G, and let $s - \alpha$ be a factor of $|T|$. Let $s - \alpha$ be a factor of $\psi_1, \psi_2, \ldots, \psi_k$, but not of ψ_{k+1}. Then $s - \alpha$ is not a factor of $\varepsilon_1\varepsilon_2 \ldots \varepsilon_k$, because each pair ε_i, ψ_i is relatively prime. By Corollary 1 of Theorem 4.2, the least common denominator of all $k \times k$ minors of G contains $s - \alpha$ raised to the same power as it has in $|T|$. This holds for each factor of $|T|$, with the appropriate k in each case, whence $|T|$ divides ϕ. By equation (1.40) of Chapter 2 it follows conversely that ϕ divides $|T|$. Hence the result.

Example 5.1 The minors of $G(s)$ given in (4.10) are

$$k = 1: \frac{1}{(s+1)^2}, \frac{1}{(s+1)(s+2)}, \frac{1}{(s+1)(s+2)}, \frac{s+3}{(s+2)^2} \tag{5.1}$$

$$k = 2: \frac{1}{(s+1)^2(s+2)} \tag{5.2}$$

and the least common denominator of these is $(s + 1)^2(s + 2)^2$. Accordingly $\nu(G) = 4$, agreeing with (4.25).

Exercise 5.1 If G is $m \times m$ and G^{-1} exists, show that $\nu(G^{-1})$ is the degree of $\varepsilon_1\varepsilon_2 \ldots \varepsilon_m$. [Use equation (1.43) of Chapter 2 and adapt the proof of Algorithm 5.1. Alternatively, write $G = T^{-1}U$, where T and U are relatively (left) prime, and hence construct a system matrix of least order for G^{-1}.]

The above method is useful when l or m is small. When l and m are both large, the number of minors to be examined becomes excessive, and the following methods will require less work.

Algorithm 5.2 Write

$$G(s) = N(s)/d(s) + D(s) \tag{5.3}$$

where $d(s)$ is the monic least common denominator of the elements of G, $N(s)/d(s)$ is proper, and $D(s)$ is polynomial. Write also

$$d(s) = s^p + a_{p-1}s^{p-1} + \cdots + a_0 \tag{5.4}$$

$$N(s) = N_{p-1}s^{p-1} + N_{p-2}s^{p-2} + \cdots + N_0 \tag{5.5}$$

where $N_0, N_1, \ldots, N_{p-1}$ are real $m \times l$ matrices. Generate matrices Q_{ij} in the following way:

$$\left.\begin{aligned} &Q_{10} = 0, \qquad Q_{1j} = N_{j-1}, \quad j = 1, 2, \ldots, p \\ &Q_{i0} = 0, \qquad Q_{ij} = Q_{i-1,j-1} - a_{j-1}Q_{i-1,p} \\ &\qquad\qquad\qquad\qquad i = 2, 3, \ldots, p, \quad j = 1, 2, \ldots, p \end{aligned}\right\} \tag{5.6}$$

Then $\nu(G)$ is the rank of the $pm \times pl$ matrix R,

$$R = \begin{pmatrix} Q_{11} & Q_{12} & \cdots & Q_{1p} \\ Q_{21} & Q_{22} & \cdots & Q_{2p} \\ \vdots & \vdots & & \vdots \\ Q_{p1} & Q_{p2} & \cdots & Q_{pp} \end{pmatrix} \tag{5.7}$$

Proof It is easily confirmed that the system matrix

$$\begin{pmatrix} sI_{pl} - A & B \\ -C & D(s) \end{pmatrix}$$

$$= \left(\begin{array}{cccccc|c} sI_l & -I_l & 0 & \cdots & 0 & 0 & 0 \\ 0 & sI_l & -I_l & \cdots & 0 & 0 & 0 \\ \vdots & \vdots & \vdots & & \vdots & \vdots & \vdots \\ 0 & 0 & 0 & \cdots & sI_l & -I_l & 0 \\ a_0I_l & a_1I_l & a_2I_l & \cdots & a_{p-2}I_l & (s + a_{p-1})I_l & I_l \\ \hline -N_0 & -N_1 & -N_2 & \cdots & -N_{p-2} & -N_{p-1} & D(s) \end{array}\right) \tag{5.8}$$

has no i.d. zeros and gives rise to G. By system similarity it is easy to arrange A as a block diagonal matrix of identical companion matrices, each with characteristic polynomial $s^p + a_{p-1}s^{p-1} + \cdots + a_0$. This polynomial is therefore also the minimal polynomial of A. The number of o.d. zeros of the matrix (5.8) is equal, by Theorem 8.1 of Chapter 2, to the rank defect of the matrix,

$$\begin{pmatrix} C \\ CA \\ \vdots \\ CA^{p-1} \end{pmatrix} \tag{5.9}$$

Hence the order of the system resulting after the o.d. zeros are removed is the rank of the matrix (5.9). This matrix, however, is R.

EXERCISE 5.2 Prove that the system matrix in (5.8) has no i.d. zeros and gives rise to G. [Add s times (block) column p to (block) column $p - 1$, etc.]

EXERCISE 5.3 Show that if

$$G(s) = \frac{s+1}{2s^3 + 3s^2 + 2s + 1} = \frac{\frac{1}{2}s + \frac{1}{2}}{s^3 + \frac{3}{2}s^2 + s + \frac{1}{2}} \tag{5.10}$$

then

$$4R = \begin{pmatrix} 2 & 2 & 0 \\ 0 & 2 & 2 \\ -1 & -2 & -1 \end{pmatrix} \tag{5.11}$$

which has rank 2. Confirm that a factor $s + 1$ cancels from numerator and denominator in $G(s)$.

EXERCISE 5.4 With the notation of (5.4), (5.5), show that the rank of the $[(p-1)l + pm] \times (2p-1)l$ matrix R_1 is $\nu(G) + (p-1)l$, where

$$R_1 = \begin{pmatrix} I_l & a_{p-1}I_l & a_{p-2}I_l & \cdots & a_2I_l & a_1I_l & a_0I_l & 0 & \cdots & 0 \\ 0 & I_l & a_{p-1}I_l & \cdots & a_3I_l & a_2I_l & a_1I_l & a_0I_l & \cdots & 0 \\ \vdots & \vdots & \vdots & & \vdots & \vdots & \vdots & \vdots & & \vdots \\ 0 & 0 & 0 & \cdots & I_l & a_{p-1}I_l & a_{p-2}I_l & a_{p-3}I_l & \cdots & a_0I_l \\ 0 & 0 & 0 & \cdots & 0 & N_{p-1} & N_{p-2} & N_{p-3} & \cdots & N_0 \\ 0 & 0 & 0 & \cdots & N_{p-1} & N_{p-2} & N_{p-3} & N_{p-4} & \cdots & 0 \\ \vdots & \vdots & \vdots & & \vdots & \vdots & \vdots & \vdots & & \vdots \\ N_{p-1} & N_{p-2} & N_{p-3} & \cdots & N_1 & N_0 & 0 & 0 & \cdots & 0 \end{pmatrix} \tag{5.12}$$

[By (block) row operations bring the last p (block) rows to the form $(0, R)$, where R is given by (5.7). The matrix R_1 is a generalized resultant matrix.]

EXERCISE 5.5 If $G(s) = N_0/d(s)$, show that

$$R = \begin{pmatrix} N_0 & 0 & \dots & 0 \\ 0 & N_0 & \dots & 0 \\ \vdots & \vdots & & \vdots \\ 0 & 0 & \dots & N_0 \end{pmatrix} \tag{5.13}$$

and so $\nu(G) = p$ rank N_0.

Algorithm 5.3 Expand $G(s)$ element by element in a Laurent series, convergent for large enough $|s|$,

$$G(s) = D_q s^q + D_{q-1}s^{q-1} + \cdots + D_0 + G_1 s^{-1} + G_2 s^{-2} + \cdots \tag{5.14}$$

Then with p defined as in (5.4), $\nu(G)$ is the rank of the $mp \times lp$ matrix H,

$$H = \begin{pmatrix} G_1 & G_2 & \dots & G_p \\ G_2 & G_3 & \dots & G_{p+1} \\ \vdots & \vdots & & \vdots \\ G_p & G_{p+1} & \dots & G_{2p-1} \end{pmatrix} \tag{5.15}$$

Proof Let a system matrix of least order giving rise to G be

$$\begin{pmatrix} sI_\nu - A & B \\ -C & D(s) \end{pmatrix} \tag{5.16}$$

Then because $(sI_\nu - A)^{-1}$ can be expanded in the Laurent series, valid for large enough $|s|$,

$$(sI_\nu - A)^{-1} = s^{-1}I_\nu + s^{-2}A + s^{-3}A^2 + \cdots \tag{5.17}$$

it follows that

$$H = \begin{pmatrix} C \\ CA \\ \vdots \\ CA^{p-1} \end{pmatrix} (B \quad AB \quad \dots \quad A^{p-1}B) \tag{5.18}$$

Because the system matrix (5.16) has least order, and because the minimal polynomial of A has degree p by Corollary 3 of Theorem 4.2, each factor in (5.18) has rank ν. Hence (Chapter 1, Section 1) H has rank ν.

EXERCISE 5.6 For the $G(s)$ given in (5.10), show that

$$16H = \begin{pmatrix} 0 & 8 & -4 \\ 8 & -4 & -2 \\ -4 & -2 & 2 \end{pmatrix} \tag{5.19}$$

which has rank 2.

EXERCISE 5.7 If $G(s) = N_0/d(s)$, as in Exercise 5.5, show that

$$H = \begin{pmatrix} 0 & 0 & 0 & \dots & N_0 \\ \vdots & \vdots & \vdots & & \vdots \\ 0 & 0 & N_0 & \dots & G_{2p-3} \\ 0 & N_0 & -a_{p-1}N_0 & \dots & G_{2p-2} \\ N_0 & -a_{p-1}N_0 & (a_{p-1}^2 - a_{p-2})N_0 & \dots & G_{2p-1} \end{pmatrix} \tag{5.20}$$

where $G_{2p-1}, G_{2p-2}, \dots,$ contain N_0 multiplied by coefficients depending on the a_i. [Transform H into R by elementary operations.]

EXERCISE 5.8 Find $\nu(G)$ for the G of equation (4.10) by Algorithms (5.2) and (5.3).

The relation between R and H is pursued in the Problems at the end of this chapter.

5.2 Generation of system matrices from G

If G is given, and we wish to generate a P which gives rise to G, many routes are open to us. In the first place, G is already a rational system matrix, and it can be transformed by system equivalence into any other P giving rise to G. Often, however, P will be required to have polynomial, or state-space, form and to have least order. It is then easier to write down a system matrix in the required form (either polynomial or state-space) and remove any decoupling zeros which it may have. Once a system matrix of least order is obtained, it may be transformed as required by strict system equivalence or system similarity.

Example 5.2 A polynomial system matrix giving rise to the G of equation (4.10) is

$$\left(\begin{array}{cccc|cc} I_4 & 0 & 0 & & 0 & 0 \\ 0 & (s+1)^2(s+2) & 0 & & s+2 & s+1 \\ 0 & 0 & (s+1)(s+2)^2 & & s+2 & (s+1)(s+3) \\ \hline 0 & -1 & 0 & & 0 & 0 \\ 0 & 0 & -1 & & 0 & 0 \end{array}\right) \tag{5.21}$$

which has no o.d. zeros. On putting $s = -2$ it is clear that rows 5 and 6 are linearly dependent, and subtracting one from the other will give a zero row. So subtracting row 5 from row 6 in (5.21), dividing through by $s + 2$, and reducing I_4 to I_3, we have

$$\left(\begin{array}{ccc|cc} I_3 & 0 & 0 & 0 & 0 \\ 0 & (s+1)^2(s+2) & 0 & s+2 & s+1 \\ 0 & -(s+1)^2 & (s+1)(s+2) & 0 & s+1 \\ \hline 0 & -1 & 0 & 0 & 0 \\ 0 & 0 & -1 & 0 & 0 \end{array}\right) \tag{5.22}$$

Without further work, a factor $s + 1$ can be divided out from row 5, and after reducing I_3 to I_2 this gives

$$\left(\begin{array}{ccc|cc} I_2 & 0 & 0 & 0 & 0 \\ 0 & (s+1)^2(s+2) & 0 & s+2 & s+1 \\ 0 & -(s+1) & (s+2) & 0 & 1 \\ \hline 0 & -1 & 0 & 0 & 0 \\ 0 & 0 & -1 & 0 & 0 \end{array}\right) \tag{5.23}$$

The only possible i.d. zeros in (5.23) are the zeros -1, -2 of $|T|$. It is therefore easy to see that (5.23) has no i.d. zeros, and so has least order.

If a system matrix of least order in state-space form is required, it can be obtained by first forming a polynomial system matrix of least order, as in Example 5.2, and then bringing this to state-space form by strict system equivalence. A more convenient procedure, however, will usually be the following. If G is not proper, put $G(s) = G_0(s) + D(s)$, where $D(s)$ is a polynomial matrix and G_0 is proper. Let $d_i(s)$ be the least common denominator of the ith row of G_0, and write G_0 in terms of its rows,

$$G_0(s) = \begin{pmatrix} u_1(s)/d_1(s) \\ u_2(s)/d_2(s) \\ \vdots \\ u_m(s)/d_m(s) \end{pmatrix} \tag{5.24}$$

Write also

$$d_i(s) = s^{p_i} + a^i_{p_i-1}s^{p_i-1} + \cdots + a^i_0 \tag{5.25}$$

$$u_i(s) = u^i_{p_i-1}s^{p_i-1} + u^i_{p_i-2}s^{p_i-2} + \cdots + u^i_0 \tag{5.26}$$

Then a system matrix in state-space form giving rise to G is

$$P(s) = \left(\begin{array}{cccc|c} sI_{p_1} - A_1 & 0 & \dots & 0 & B_1 \\ 0 & sI_{p_2} - A_2 & \dots & 0 & B_2 \\ \vdots & \vdots & & \vdots & \vdots \\ 0 & 0 & \dots & sI_{p_m} - A_m & B_m \\ \hline -C_1 & -C_2 & \dots & -C_m & D(s) \end{array}\right) \tag{5.27}$$

in which the A_i are companion matrices,

$$A_i = \begin{pmatrix} 0 & 0 & \dots & 0 & 0 & -a_0^i \\ 1 & 0 & \dots & 0 & 0 & -a_1^i \\ \vdots & \vdots & & \vdots & \vdots & \vdots \\ 0 & 0 & \dots & 1 & 0 & -a_{p_i-2}^i \\ 0 & 0 & \dots & 0 & 1 & -a_{p_i-1}^i \end{pmatrix} \tag{5.28}$$

while the B_i and C_i are respectively,

$$B_i = \begin{pmatrix} u_0 \\ u_1^i \\ \vdots \\ u_{p_i-1}^i \end{pmatrix}, \qquad C_i = (0 \quad 0 \quad \dots \quad 0 \quad e_i) \tag{5.29}$$

and e_i is the ith column of I_m. The matrix P has no o.d. zeros, as can readily be confirmed. It may have i.d. zeros, and if so these can be removed by Algorithm 7.1 of Chapter 2. A similar development of G in terms of columns instead of rows is also possible; it may have an advantage depending on the relative degrees of common denominators for rows and for columns.

EXERCISE 5.9 Confirm that P given by (5.27) gives rise to G and has no o.d. zeros. [Compare the procedure in Section 2.]

Example 5.3 A system matrix in state-space form giving rise to the G of equation (4.10) is [compare (5.21)]

$$\left(\begin{array}{cccccc|cc} s & 0 & 2 & 0 & 0 & 0 & 2 & 1 \\ -1 & s & 5 & 0 & 0 & 0 & 1 & 1 \\ 0 & -1 & s+4 & 0 & 0 & 0 & 0 & 0 \\ 0 & 0 & 0 & s & 0 & 4 & 2 & 3 \\ 0 & 0 & 0 & -1 & s & 8 & 1 & 4 \\ 0 & 0 & 0 & 0 & -1 & s+5 & 0 & 1 \\ \hline 0 & 0 & -1 & 0 & 0 & 0 & 0 & 0 \\ 0 & 0 & 0 & 0 & 0 & -1 & 0 & 0 \end{array}\right) \tag{5.30}$$

where two companion matrices have been written, with characteristic polynomials respectively,

$$(s + 1)^2(s + 2) = s^3 + 4s^2 + 5s + 2 \tag{5.31}$$

$$(s + 1)(s + 2)^2 = s^3 + 5s^2 + 8s + 4 \tag{5.32}$$

The columns of B were obtained from the coefficients of successive powers of s in $U(s)$ of (5.21). Then by the procedure of Chapter 2, Section 7, the matrix (5.30) can be brought to the form

$$\left(\begin{array}{cccccc:cc} s & 2 & 0 & 0 & 0 & 0 & 0 & 0 \\ -1 & s+3 & 0 & 0 & 0 & 0 & 0 & 0 \\ 0 & -1 & s+1 & -1 & 0 & 0 & 0 & 0 \\ 0 & 0 & 0 & s+2 & -1 & -1 & 0 & 0 \\ 0 & 0 & 0 & -1 & s+2 & 1 & 1 & 0 \\ 0 & 0 & 0 & 0 & -1 & s+1 & 0 & 1 \\ \hdashline 0 & 0 & -1 & -1 & 0 & 0 & 0 & 0 \\ 0 & 0 & 0 & 0 & 0 & -1 & 0 & 0 \end{array}\right) \tag{5.33}$$

Deleting the first two rows and first two columns leaves a state-space representation of least order.

5.3 Summary of transformations

The results which have been obtained on the transformation of rational, polynomial, and state-space system matrices giving rise to the same G are summarized in Fig. 1 in which it is assumed that $r = n$. Given G, which is a rational system matrix, it can be transformed by system equivalence to any other system matrix giving rise to G. A subset of these system matrices is the set of polynomial system matrices, which have an order n associated with them. All polynomial system matrices with least order $\nu(G)$ are related by strict system equivalence. Polynomial system matrices of greater than least order may or may not be related by strict system equivalence: necessary and sufficient conditions for them to be so related are not known. If P_1 and P_2 with order $n > \nu(g)$ are related by strict system equivalence, then they have in common the number, and values of, their i.d., o.d., and i.o.d. zeros.

A subset of the polynomial system matrices is the set of state-space matrices. If these have least order they are related by system similarity. If they do not have least order they are system similar if and only if they are strictly system equivalent.

Passage from the plane of order $n > \nu(g)$ to the plane of order $\nu(G)$ is by system equivalence, which deletes the decoupling zeros. Passage in the reverse direction can be made by a special form of system equivalence in which the transforming matrices M, N, X, Y are polynomial [Chapter 2, equation (6.3)]. If we allow $r > n$ some obvious changes are needed in Fig. 1.

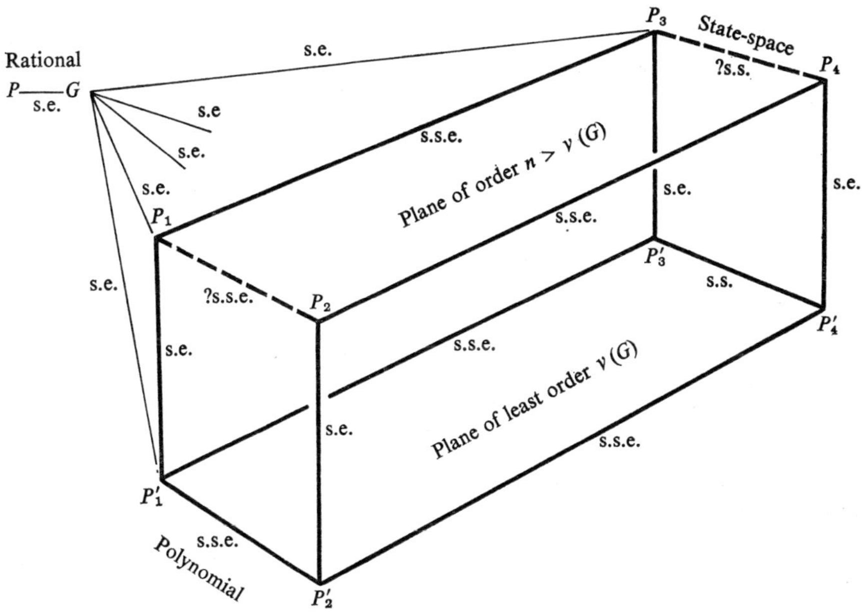

FIG. 1 Transformations of system matrices when $r = n$

6 Series and parallel connection of systems

Suppose that two polynomial system matrices P_1, P_2 are given, each having least order, and giving rise respectively to transfer function matrices G_1, G_2. Suppose further that G_1, G_2 are compatible for the product G_2G_1, and that the systems represented by P_1 and P_2 are connected in series to give the transfer function G_2G_1. Two interesting questions arise:

(i) Does the cascaded system have least order?
(ii) Does the answer to (i) depend on the systems P_1, P_2, or only on G_1 and G_2?

If G_1 and G_2 are compatible for the sum G_1 and G_2, similar questions can be asked about the parallel connection of the systems represented by P_1 and P_2.

Throughout this section we shall use 'pole of G' to mean a zero (necessarily finite) of one of the denominator polynomials $\psi_i(s)$ in the McMillan form M of

G. Similarly 'zero of G' will mean a finite zero of one of the numerator polynomials $\varepsilon_i(s)$ of M. These are the meanings assigned in Section 4.

6.1 Systems in series

Let the equations for the two systems be

$$\left.\begin{aligned} T_1\bar{\xi}_1 &= U_1\bar{u}_1 \\ \bar{y}_1 &= V_1\bar{\xi}_1 + W_1\bar{u}_1 \end{aligned}\right\} \tag{6.1}$$

$$\left.\begin{aligned} T_2\bar{\xi}_2 &= U_2\bar{u}_2 \\ \bar{y}_2 &= V_2\bar{\xi}_2 + W_2\bar{u}_2 \end{aligned}\right\} \tag{6.2}$$

where $\bar{u}_1$, $\bar{y}_1$, $\bar{u}_2$, $\bar{y}_2$, $\bar{\xi}_1$, $\bar{\xi}_2$ are respectively of dimension l, k, k, m, r_1, r_2, and set

$$\bar{u}_2 = \bar{y}_1 \tag{6.3}$$

These equations can be written

$$\left(\begin{array}{ccc|c} 0 & T_1 & 0 & U_1 \\ T_2 & 0 & U_2 & 0 \\ 0 & -V_1 & -I_k & W_1 \\ \hline -V_2 & 0 & W_2 & 0 \end{array}\right)\left(\begin{array}{c} \bar{\xi}_2 \\ \bar{\xi}_1 \\ -\bar{u}_2 \\ \hline -\bar{u}_1 \end{array}\right) = \left(\begin{array}{c} 0 \\ 0 \\ 0 \\ \hline -\bar{y}_2 \end{array}\right) \tag{6.4}$$

where the first matrix, say P, qualifies as a polynomial system matrix for the complete system provided (as we assume) that the matrices T_1 and T_2 have dimension at least equal to the degrees of $|T_1|$ and $|T_2|$. For it follows from

$$\begin{vmatrix} 0 & T_1 & 0 \\ T_2 & 0 & U_2 \\ 0 & -V_1 & -I_k \end{vmatrix} = |T_1|\,|T_2| \tag{6.5}$$

that a similar condition is satisfied for the leading 3×3 (block) matrix of P.

Now let the systems (6.1) and (6.2) be transformed by strict system equivalence, and distinguish the transforming matrices in an obvious way as M_1, M_2, etc. It is easy to verify that P becomes

$$\left(\begin{array}{ccc|c} 0 & M_1T_1N_1 & 0 & M_1T_1Y_1+M_1U_1 \\ M_2T_2N_2 & 0 & M_2T_2Y_2+M_2U_2 & 0 \\ 0 & X_1T_1N_1-V_1N_1 & -I_k & \begin{array}{r} X_1T_1Y_1-V_1Y_1 \\ +X_1U_1+W_1 \end{array} \\ \hline X_2T_2N_2-V_2N_2 & 0 & \begin{array}{r} X_2T_2Y_2-V_2Y_2 \\ +X_2U_2+W_2 \end{array} & 0 \end{array}\right) \tag{6.6}$$

This is related to P by a transformation of strict system equivalence, with transforming matrices

$$\left(\begin{array}{ccc|c} M_1 & 0 & 0 & 0 \\ 0 & M_2 & 0 & 0 \\ X_1 & 0 & I_k & 0 \\ \hline 0 & X_2 & 0 & I_m \end{array}\right) \tag{6.7}$$

and

$$\left(\begin{array}{ccc|c} N_2 & 0 & Y_2 & 0 \\ 0 & N_1 & 0 & Y_1 \\ 0 & 0 & I_k & 0 \\ \hline 0 & 0 & 0 & I_l \end{array}\right) \tag{6.8}$$

The i.d. zeros, o.d. zeros, and i.o.d. zeros of the cascaded combination are those of P, and they are invariant under pre- and post-multiplication by (6.7) and (6.8) respectively. But by Theorem 3.1, all P_1 and P_2 of least order giving rise to G_1 and G_2 respectively are related as in (6.6). It follows that the i.d., o.d., and i.o.d. zeros of P are invariant under the permitted choice of P_1 and P_2.

A simple way to find the decoupling zeros is to express G_1 and G_2 in the standard forms,

$$G_2 = V_2T_2^{-1} + D_2 = (V_2 + D_2T_2)T_2^{-1} \tag{6.9}$$

$$G_1 = T_1^{-1}U_1 + D_1 = T_1^{-1}(U_1 + T_1D_1) \tag{6.10}$$

where T_1, T_2, U_1, V_2 are no longer the same as before. Then it is easily seen that the combined system has the equations (in which $\bar{\xi}_2$ is no longer the same as before)

$$\left.\begin{aligned} T_1T_2\bar{\xi}_2 &= (U_1 + T_1D_1)\bar{u}_1 \\ \bar{y}_2 &= (V_2 + D_2T_2)\bar{\xi}_2 \end{aligned}\right\} \tag{6.11}$$

to which corresponds the system matrix

$$\left(\begin{array}{cc|c} I_{r_1+r_2-k} & 0 & 0 \\ 0 & T_1T_2 & U_1 + T_1D_1 \\ \hline 0 & -(V_2 + D_2T_2) & 0 \end{array}\right) \tag{6.12}$$

The decoupling zeros of the cascaded system are those of (6.12).

Example 6.1 Let

$$G_1(s) = \begin{pmatrix} \dfrac{s+2}{(s+1)^2} & 0 \\ 0 & \dfrac{s+1}{(s+2)^2} \end{pmatrix} = \begin{pmatrix} (s+1)^2 & 0 \\ 0 & (s+2)^2 \end{pmatrix}^{-1} \begin{pmatrix} s+2 & 0 \\ 0 & s+1 \end{pmatrix} \tag{6.13}$$

$$G_2(s) = \begin{pmatrix} \dfrac{s+1}{(s+2)^2} & 0 \\ \dfrac{-1}{(s+1)(s+2)} & \dfrac{s+2}{(s+1)^2} \end{pmatrix}$$
$$= \begin{pmatrix} s+1 & 0 \\ 0 & s+2 \end{pmatrix} \begin{pmatrix} (s+2)^2 & 0 \\ s+1 & (s+1)^2 \end{pmatrix}^{-1} \tag{6.14}$$

so that a system matrix for the combined system is

$$\left(\begin{array}{ccc|cc} I_6 & 0 & 0 & 0 & 0 \\ 0 & (s+1)^2(s+2)^2 & 0 & s+2 & 0 \\ 0 & (s+1)(s+2)^2 & (s+1)^2(s+2)^2 & 0 & s+1 \\ \hline 0 & -(s+1) & 0 & 0 & 0 \\ 0 & 0 & -(s+2) & 0 & 0 \end{array}\right) \tag{6.15}$$

The i.d. zeros are -1, -2 and the o.d. zeros are -1, -2. However, when the i.d. zeros have been removed there remains only the one o.d. zero -2. There is, therefore, one i.o.d. zero, -1. The least order $\nu(G_2G_1)$ is $8 - 2 - 2 + 1 = 5$, and the poles of G_2G_1 are given as in Chapter 2, equation (5.12) by

$$\begin{aligned} \{\alpha_i\} &= \{-1, -1, -1, -1, -2, -2, -2, -2\} - \{-1, -2, -1, -2\} + \{-1\} \\ &= \{-1, -1, -1, -2, -2\} \end{aligned} \tag{6.16}$$

In fact the combined transfer function is

$$G_2G_1 = \begin{pmatrix} \dfrac{1}{(s+1)(s+2)} & 0 \\ \dfrac{-1}{(s+1)^3} & \dfrac{1}{(s+1)(s+2)} \end{pmatrix} \tag{6.17}$$

which has McMillan form

$$\begin{pmatrix} \dfrac{1}{(s+1)^3(s+2)} & 0 \\ 0 & \dfrac{s+1}{s+2} \end{pmatrix} \tag{6.18}$$

agreeing with (6.16). By Chapter 2, Section 7, these results mean that a state-space representation of the two cascaded least-order systems would have:

a controllable and observable subspace of dimension 5;
a controllable but unobservable subspace of dimension 1;
an uncontrollable but observable subspace of dimension 1;
an uncontrollable and unobservable subspace of dimension 1.

Theorem 6.1 Let G_2 be $m \times k$ and G_1 be $k \times l$. Then every i.d. zero of the cascaded system corresponding to G_2G_1 is a pole of G_2 (and if $l \geqslant k$ is a zero of G_1). Every o.d. zero of the cascaded system is a pole of G_1 (and if $m \geqslant k$ is a zero of G_2).

Proof In (6.12), let the $k \times (k + l)$ matrix

$$\begin{pmatrix} T_1(\beta)T_2(\beta) & U_1(\beta) + T_1(\beta)D_1(\beta) \end{pmatrix} \tag{6.19}$$

have rank less than k. It follows that $U_1(\beta) + T_1(\beta)D_1(\beta)$ has rank less than k, and if $l \geqslant k$ then β is a zero of G_1 by Theorem 4.3. Also

$$\begin{pmatrix} T_1(\beta)T_2(\beta) & U_1(\beta) + T_1(\beta)D_1(\beta) \end{pmatrix} = \begin{pmatrix} T_1(\beta) & U_1(\beta) \end{pmatrix}\begin{pmatrix} T_2(\beta) & D_1(\beta) \\ 0 & I_l \end{pmatrix} \tag{6.20}$$

and $\begin{pmatrix} T_1(s) & U_1(s) \end{pmatrix}$ has rank k for all s. It follows that $T_2(\beta)$ is singular and β is a pole of G_2. The second part follows in the same way.

It should be noticed that by 'β is a zero of G_1' we mean that in the McMillan form of G_1 at least one of the $\varepsilon_i(\beta)$ is zero. If one of the ε_i is identically zero, then any finite β is a zero of G_1.

EXERCISE 6.1 Investigate the decoupling zeros of a cascaded system corresponding to G_2G_1, where

$$G_1 = \begin{pmatrix} \dfrac{s+1}{(s+3)^2} \\ \dfrac{s+2}{(s+3)^2} \end{pmatrix} \tag{6.21}$$

$$G_2 = \begin{pmatrix} \dfrac{1}{s+1} & \dfrac{-1}{s+2} \\ \dfrac{1}{s+1} & \dfrac{1}{s+4} \end{pmatrix} \tag{6.22}$$

Theorem 6.2 Let G_2 be $m \times k$ and G_1 be $k \times l$, with $m \geqslant k$, $l \geqslant k$. If no zero of G_1 is a pole of G_2, and no zero of G_2 is a pole of G_1, then the cascaded system has least order, and

(i) the zeros of G_2G_1 are those of G_1 together with those of G_2;
(ii) the poles of G_2G_1 are those of G_1 together with those of G_2.

Proof Under the conditions of the theorem, the system matrix (6.12) has no decoupling zeros by Theorem 6.1 By Theorem 4.1 the poles of G_2G_1 are then the zeros of $|T_1T_2|$. The zeros of G_2G_1 are the zeros of

$$\begin{vmatrix} T_1T_2 & U_1 + T_1D_1 \\ -(V_2 + D_2T_2) & 0 \end{vmatrix} = |U_1 + T_1D_1|\,|V_2 + D_2T_2| \tag{6.23}$$

The theorem then follows from Theorem 4.3.

When $k = l = m = 1$, the converse of Theorem 6.2 holds. In the general case, however, the converse is not true, as is shown by the following example.

Example 6.2 Let

$$G_1 = \begin{pmatrix} \dfrac{s+1}{(s+2)^2} & 0 \\ 0 & \dfrac{s+3}{(s+4)^2} \end{pmatrix}, \qquad G_2 = \begin{pmatrix} \dfrac{s+4}{(s+3)^2} & 0 \\ 0 & \dfrac{s+2}{(s+1)^2} \end{pmatrix} \tag{6.24}$$

so that the McMillan forms are

$$\left.\begin{aligned} M_1 &= \begin{pmatrix} \dfrac{1}{(s+2)^2(s+4)^2} & 0 \\ 0 & (s+1)(s+3) \end{pmatrix} \\ M_2 &= \begin{pmatrix} \dfrac{1}{(s+1)^2(s+3)^2} & 0 \\ 0 & (s+2)(s+4) \end{pmatrix} \end{aligned}\right\} \tag{6.25}$$

The conditions of Theorem 6.2 are not fulfilled, yet

$$G_2G_1 = \begin{pmatrix} \dfrac{(s+1)(s+4)}{(s+2)^2(s+3)^2} & 0 \\ 0 & \dfrac{(s+2)(s+3)}{(s+1)^2(s+4)^2} \end{pmatrix} \tag{6.26}$$

with McMillan form

$$\begin{pmatrix} \dfrac{1}{(s+1)^2(s+2)^2(s+3)^2(s+4)^2} & 0 \\ 0 & (s+1)(s+2)(s+3)(s+4) \end{pmatrix} \tag{6.27}$$

so that the conclusions of the theorem hold.

Theorem 6.3 If $G_1, G_2, \ldots, G_p$ are $k \times k$ matrices and no pole of any G_i is a zero of any other G_j, then the cascaded system corresponding to $G_pG_{p-1}\ldots G_1$ has least order.

Proof Theorem 6.1 shows that the cascaded system corresponding to G_2G_1 has least order. By Theorem 6.2 and another application of Theorem 6.1, the cascaded system corresponding to $G_3(G_2G_1)$ has least order. In this way the theorem is established.

6.2 Systems in parallel

The behaviour of systems connected in parallel is somewhat easier to investigate than when they are connected in series. If the two least-order systems (6.1) and (6.2) are compatible for the sum $G_1 + G_2$, and if

$$\bar{u}_1 = \bar{u}_2 \tag{6.28}$$

$$\bar{y} = \bar{y}_1 + \bar{y}_2 \tag{6.29}$$

then a system matrix for the parallel combination is

$$\left(\begin{array}{cc|c} T_1 & 0 & U_1 \\ 0 & T_2 & U_2 \\ \hline -V_1 & -V_2 & W_1 + W_2 \end{array}\right) \tag{6.30}$$

EXERCISE 6.2 Show that the i.d., o.d., and i.o.d. zeros of the parallel system depend only on G_1 and G_2, and not on the particular (least-order) systems by which they are realized.

Theorem 6.4 If no pole of G_1 is a pole of G_2, then the parallel system has least order and its poles are the poles of G_1 together with the poles of G_2.

Proof In (6.30) the rank of the matrix

$$\begin{pmatrix} T_1 & 0 & U_1 \\ 0 & T_2 & U_2 \end{pmatrix} \tag{6.31}$$

can be less than its full value only if $|T_1T_2| = 0$. If $|T_1(\eta)| = 0$ then the rows of $(T_1(\eta)\ 0\ U_1(\eta))$ are linearly independent because the system (6.1) has least order. There is therefore a set of columns of $(T_1(\eta)\ 0\ U_1(\eta))$ having non-zero determinant. Also $|T_2(\eta)| \neq 0$ by the assumption of the theorem, and so there is a non-zero minor of maximal size in (6.31). Hence the system matrix (6.30) has no i.d. zeros, and similarly it has no o.d. zeros. The poles of the transfer function to which (6.30) gives rise are the zeros of $|T_1T_2|$, which completes the proof.

When G_1 and G_2 are both 1×1, the converse of Theorem 6.4 is true, but this is no longer so in the general case.

EXERCISE 6.3 Show that with G_1, G_2 as below, the conclusions of Theorem 6.4 are true although the condition in the theorem is not satisfied.

$$G_1 = \begin{pmatrix} \frac{0}{s+1} & 0 \\ 0 & \frac{1}{s+2} \end{pmatrix}, \qquad G_2 = \begin{pmatrix} \frac{1}{s+2} & 0 \\ 0 & \frac{1}{s+1} \end{pmatrix} \tag{6.32}$$

7 Degree of a rational matrix

Though the degree of a polynomial is defined, no attempt has been made so far to define the degree of a polynomial matrix. The reason is that we wish to define a degree for rational matrices. As polynomial matrices are a special case of rational matrices, their degree must be consistent with the more general definition.

If two polynomials $u(s)$ and $t(s)$ are given, the degree $\delta(g)$ of $g(s) = u(s)/t(s)$ is defined as the greater of the degree $\delta(u)$ of $u(s)$ and the degree $\delta(t)$ of $t(s)$. Alternatively we may define $\delta(g)$ as the sum of the orders of the poles of $g(s)$, including the pole at infinity. It is easy to see that the two definitions agree, and that $\delta(g)$ so defined has the following properties:

(i) $\delta(g) \geqslant 0$ and is an integer.
(ii) $\delta(g) = 0$ implies that g is independent of s.
(iii) $\delta(g^{-1}) = \delta(g)$.
(iv) $\delta(g_1) + \delta(g_2) \geqslant \delta(g_1 + g_2)$, with equality holding if g_1 and g_2 have no (finite or infinite) pole in common.
(v) $\delta(g_1) + \delta(g_2) \geqslant \delta(g_1 g_2)$, with equality holding if no (finite or infinite) pole of g_1 is a zero of g_2, and no (finite or infinite) pole of g_2 is a zero of g_1.

Our aim is to find a definition for $\delta(G)$, where G is a rational matrix, which possesses similar properties.

In this section 'pole of G' and 'zero of G' will be used with a wider meaning than in the previous section. The finite poles and zeros of G will be defined as before. If any element of G tends to infinity as $s \to \infty$, G will be said to have a pole at infinity. If every minor of some given order k of G tends to zero as $s \to \infty$, G will be said to have a zero at infinity.

It has been seen in Section 4.1 that a rational matrix G can be written

$$G(s) = T^{-1}(s)U(s) + D(s) \tag{7.1}$$

$$= T^{-1}(s)[U(s) + T(s)D(s)] \tag{7.2}$$

where T and U, and therefore also T and $U + TD$, are relatively (left) prime. If G is square, a candidate for $\delta(G)$ might appear to be $\delta(g)$, where

$$g(s) = \frac{|U(s) + T(s)D(s)|}{|T(s)|} \tag{7.3}$$

and where $\delta(g)$ is defined in an earlier paragraph. Such a definition is unsatisfactory, however, because there is an inherent asymmetry in the treatment of finite and infinite poles.

Example 7.1 If

$$G(s) = \begin{pmatrix} 1 & 1/s \\ 0 & 1 \end{pmatrix} \tag{7.4}$$

then

$$G(s) = T^{-1}(s)U(s) = \begin{pmatrix} s & 0 \\ 0 & 1 \end{pmatrix}^{-1} \begin{pmatrix} s & 1 \\ 0 & 1 \end{pmatrix} \tag{7.5}$$

and $\delta(|T|) = 1$, $\delta(|U|) = 1$. On the other hand if

$$G_1(s) = \begin{pmatrix} 1 & s \\ 0 & 1 \end{pmatrix} \tag{7.6}$$

then

$$G_1(s) = T_1^{-1}(s)U_1(s) = \begin{pmatrix} 1 & 0 \\ 0 & 1 \end{pmatrix}^{-1} \begin{pmatrix} 1 & s \\ 0 & 1 \end{pmatrix} \tag{7.7}$$

and $\delta(|T_1|) = 0$, $\delta(|U_1|) = 0$. To achieve symmetry we should have to write, instead of (7.7),

$$G_1(s) = \begin{pmatrix} 1/s & 0 \\ 0 & 1 \end{pmatrix}^{-1} \begin{pmatrix} 1/s & 1 \\ 0 & 1 \end{pmatrix} \tag{7.8}$$

This is the underlying idea of the treatment which follows.

To re-establish uniformity in the treatment of finite and infinite poles, a change of variable is made. Let

$$p = \frac{s}{s - \alpha} \tag{7.9}$$

where α is not equal to any finite pole or zero of a minor of any order of $G(s)$. Then

$$s = \frac{\alpha p}{p - 1} \tag{7.10}$$

and so $G(s)$ can be written

$$G(s) = G\left(\frac{\alpha p}{p-1}\right) = H(p) \tag{7.11}$$

It is easy to see that G has an infinite pole if and only if H has a pole (in the sense of Section 6) at $p = 1$. Similarly, G has an infinite zero if and only if H has a zero (in the sense of Section 6) at $p = 1$.

We shall define the *degree* (or *McMillan degree*) of G by $\delta(G) = \nu(H)$, and shall proceed to show that this definition makes $\delta(G)$ independent of the choice of α, subject to the conditions given above. It also endows $\delta(G)$ with properties similar to those listed earlier for $\delta(g)$.

Theorem 7.1 With the above definition, $\delta(G)$ is independent of the choice of α, provided that α is not equal to any finite pole or zero of a minor of any order of G.

Proof Let

$$p_1 = \frac{s}{s-\alpha_1} \tag{7.12}$$

$$p_2 = \frac{s}{s-\alpha_2} \tag{7.13}$$

where α_1 and α_2 satisfy the conditions of the theorem. Let a minor of G of some order be

$$\phi(s) = \frac{s^j + a_{j-1}s^{j-1} + \cdots + a_0}{s^k + b_{k-1}s^{k-1} + \cdots + b_0} \tag{7.14}$$

where the numerator and denominator have no common factor. The substitution $s = \alpha_1 p_1/(p_1 - 1)$ leads to

$$\psi(p_1) = \frac{c_q p_1^q + c_{q-1}p_1^{q-1} + \cdots + c_0}{d_q p_1^q + d_{q-1}p_1^{q-1} + \cdots + d_0} \tag{7.15}$$

where q is the greater of j, k. Taking finite and infinite poles together, $\phi(s)$ has q poles. Similarly it has q zeros. These poles and zeros all map into finite positions in the complex plane under the transformation (7.12), which shows that neither c_q nor d_q in (7.15) is zero. Also the numerator and denominator of $\psi(p_1)$ have no common factor. For if β were a common factor in $\psi(p_1)$, then $\alpha_1\beta/(\beta - 1)$ would be a common factor of numerator and denominator in $\phi(s)$.

In the same way, substitution from (7.13) transforms $\phi(s)$ into a rational function $\chi(p_2)$ with its numerator and denominator each of degree q and having no factor in common. Each zero γ_1 of the denominator of $\psi(p_1)$ corresponds to a zero γ_2 of the denominator of $\chi(p_2)$, where by (7.12) and (7.13)

$$\gamma_2 = \frac{\alpha_1\gamma_1}{(\alpha_1 - \alpha_2)\gamma_1 + \alpha_2} \tag{7.16}$$

From these facts it follows that the least common denominator of all the minors of all orders of $G(\alpha_1 p_1/(p_1 - 1))$ has the same degree as the corresponding least common denominator of $G(\alpha_2 p_2/(p_2 - 1))$. By Algorithm 5.1 this proves the theorem.

Example 7.2 In equation (7.4) set $s = p/(p - 1)$ to obtain

$$H(p) = \begin{pmatrix} 1 & \dfrac{p-1}{p} \\ 0 & 1 \end{pmatrix} \tag{7.17}$$

Then $\nu(H)$ is clearly 1, and so $\delta(G) = 1$. The same substitution in (7.6) gives

$$H_1(p) = \begin{pmatrix} 1 & \dfrac{p}{p-1} \\ 0 & 1 \end{pmatrix} \tag{7.18}$$

which gives $\nu(H_1) = \delta(G_1) = 1$. Therefore the definition achieves symmetry between finite and infinite poles.

EXERCISE 7.1 Show that if $D(s)$ is a polynomial matrix, $\delta(D)$ as defined above is the same as $\nu[D(s^{-1})]$. [Note that the only essential properties of the transformation (7.9) in the context of this example are that it is one-to-one, and it maps the point at infinity into a finite point.]

Theorem 7.2 The degree $\delta(G)$ of a rational $m \times l$ matrix $G(s)$ has the following properties:

(i) $\delta(G) \geqslant 0$ and is an integer.
(ii) $\delta(G) = 0$ implies that G is independent of s.
(iii) If G^{-1} exists, $\delta(G^{-1}) = \delta(G)$.
(iv) If G_1 and G_2 are conformable for addition, $\delta(G_1) + \delta(G_2) \geqslant \delta(G_1 + G_2)$, with equality holding if G_1 and G_2 have no (finite or infinite) pole in common.
(v) If G_2 and G_1 are respectively $m \times k$ and $k \times l$, $\delta(G_1) + \delta(G_2) \geqslant \delta(G_2G_1)$; with equality holding if $m \geqslant k$, $l \geqslant k$, no (finite or infinite) pole of G_1 is a zero of G_2, and no (finite or infinite) pole of G_2 is a zero of G_1.

Proof Property (i) is obvious. If $\delta(G) = 0$, then $\nu\big(H(p)\big) = 0$. Write $H(p)$ in the standard form of equation (4.34)

$$H(p) = T^{-1}(p)U(p) + D(p) \tag{7.19}$$

Then because $\nu(H) = 0$, $|T|$ is a constant, and by (iii) of Theorem (4.3), T is the identity matrix. Then by (iv) of Theorem (4.3), $U(p) \equiv 0$. It was seen in the proof of Theorem 7.1 that every element of $H(p)$ tends to a finite value as $s \to \infty$, whence $D(p)$ has only constant entries. This proves (ii).

To prove (iii), which implies $l = m$, let $d(p)$ be the monic least common denominator of the elements of $H(p)$, and write

$$H(p) = \frac{Q(p)}{d(p)} \tag{7.20}$$

Then because every minor of $H(p)$, including $|H(p)|$, tends to a finite value as $p \to \infty$,

$$\delta(|Q|) = m\ \delta(d) \tag{7.21}$$

where H is $m \times m$. If $S(p)$ is the Smith form of $Q(p)$, $\delta(|S|) = \delta(|Q|)$. After cancellation, let $S(p)/d(p)$ become diag $\big(\varepsilon_i(p)/\psi_i(p)\big)$. It follows that the degree of $\varepsilon_1\varepsilon_2 \ldots \varepsilon_m$ is the same as the degree of $\psi_1\psi_2 \ldots \psi_m$. But if the substitution (7.10) takes $G(s)$ into $H(p)$, then it takes $G^{-1}(s)$ into $H^{-1}(p)$, as can easily be seen by considering the adjoint matrices of $G(s)$ and $H(p)$. By the result of Exercise 5.1 this proves (iii).

Let α be chosen in (7.9) so that it is not equal to any finite pole or zero of a minor of any order of G_1 or G_2 in (iv). If $G = G_1 + G_2$, then

$$H(p) = G\left(\frac{\alpha p}{p - 1}\right) = G_1\left(\frac{\alpha p}{p - 1}\right) + G_2\left(\frac{\alpha p}{p - 1}\right) = H_1(p) + H_2(p) \tag{7.22}$$

Then by Theorem 6.4 applied to H_1 and H_2,

$$\nu(H) = \nu(H_1) + \nu(H_2) \tag{7.23}$$

provided that H_1 and H_2 have no pole in common. If there is such a common pole, equation (6.30) shows that

$$\nu(H) \leqslant \nu(H_1) + \nu(H_2) \tag{7.24}$$

Since H_1 and H_2 have a common pole if and only if G_1 and G_2 have a (finite or infinite) pole in common, this proves (iv).

Let α be chosen so that it is not equal to any finite pole or zero of a minor of any order of G_1 or G_2 in (v), and write $G = G_2G_1$. Then

$$H(p) = G\left(\frac{\alpha p}{p - 1}\right) = G_2\left(\frac{\alpha p}{p - 1}\right)G_1\left(\frac{\alpha p}{p - 1}\right) = H_2(p)H_1(p) \tag{7.25}$$

Substituting H_1 and H_2 for G_1 and G_2 in (6.9) and (6.10), it follows from (6.12) that

$$\nu(H) \leqslant \nu(H_1) + \nu(H_2) \tag{7.26}$$

By Theorem 6.2, equality holds in (7.36) when the conditions of (v) of this theorem are satisfied. This completes the proof.

EXERCISE 7.2 Show that if $G(s) = G_0(s) + D(s)$, where $G_0(s)$ is proper and $D(s)$ is a polynomial matrix,

$$\delta(G) = \nu\big(G_0(s)\big) + \nu\big(D(s^{-1})\big) \tag{7.27}$$

[Use Theorem 7.2(iv) and Exercise 7.1. Equation (7.27) is often used as a definition of $\delta(G)$.]

EXERCISE 7.3 Show that for the G_1 and G_2 in (6.24),

$$\delta(G_1) = 4, \qquad \delta(G_2) = 4, \qquad \delta(G_1 + G_2) = 8, \qquad \delta(G_2 G_1) = 8 \tag{7.28}$$

[Notice that the condition in (v) of Theorem 7.2 is not fulfilled. This condition, like that in (iv), is sufficient but not necessary.]

EXERCISE 7.4 Find $\delta(G)$ for the following:

$$G = \frac{s^p}{(s+1)^q} \tag{7.29}$$

$$G = \begin{pmatrix} \dfrac{1}{(s+1)^2} & 2s \\ 3 + s & 6s \end{pmatrix} \tag{7.30}$$

$$G = \begin{pmatrix} s \\ \dfrac{1}{(s+1)^2} \end{pmatrix} \tag{7.31}$$

EXERCISE 7.5 Show that if $v(s)$ is a polynomial vector, $\delta(v)$ is the highest degree of any element in v [compare Section 1.3].

For most purposes it is sufficient to consider transfer function matrices which have only finite poles. There is then no distinction between $\nu(G)$ and $\delta(G)$. In electrical network theory, however, it is sometimes convenient to idealize the behaviour of a circuit in such a way that G has a pole at infinity. In these circumstances the degree $\delta(G)$ is different from $\nu(G)$. The following theorem was proved by McMillan.

Theorem 7.3 Let $Z(s)$ be an $n \times n$ rational matrix which is symmetric, has real coefficients, and has its real part positive semidefinite for all s with $\operatorname{Re} s > 0$. Then there exists a finite passive network, a $2n$-pole, which has the impedance matrix $Z(s)$. This network can be constructed (using ideal transformers) with $\delta(Z)$ reactive elements, and no fewer.

The proof belongs in network theory and will not be given here. The fact that no fewer than $\delta(Z)$ reactive elements are needed is easy to understand when Z has only finite poles, for then $\delta(Z) = \nu(Z)$. This shows that $\delta(Z)$ is a natural extension of the idea of least order. If ideal transformers are not allowed, more than $\delta(Z)$ reactive elements may have to be used to construct the network.

Problems

1 Let v be a real n-vector and define the function f_v by

$$f_v(t) = \begin{cases} 0, & t \leqslant 0 \\ e^{At}v, & t > 0 \end{cases}$$

where A is a real $n \times n$ matrix. Show that the mapping $\phi: v \to f_v$ is linear, one-to-one, and preserves sums and scalar multiples. That is, the vector spaces V_1, V_2 over the real numbers generated respectively by the vectors v and the functions f_v are isomorphic, and so have the same dimension.

2 Show that if C is a real $m \times n$ matrix and $sI_n - A$, C are relatively (right) prime, the space V_3 generated by the functions Cf_v is isomorphic to V_2 in Problem 1. [Consider derivatives for $t > 0$.]

3 Let B be a real $n \times l$ matrix with $sI_n - A$, B relatively (left) prime and $sI_n - A$, C relatively (right) prime. Define the $r \times l$ matrix function h by

$$h(t) = \begin{cases} 0, & t \leqslant 0 \\ Ce^{At}B, & t > 0 \end{cases}$$

Let q be the degree of the minimal polynomial of A, and let V_4 be the space spanned by the columns of

$$(h, D_-h, \ldots, D_-^q h)$$

where D_-h is the left-hand derivative, $D_-^2 h = D_-(D_-h)$, etc. Show that V_4 is isomorphic to V_3 in Problem 3. [Use the fact that

$$D_-^k h(t) = \begin{cases} 0, & t \leqslant 0 \\ Ce^{At}A^kB, & t > 0] \end{cases}$$

4 Show that the Laplace transform of $D_-^k h$ in Problem 3 is the proper part of $s^k G(s)$, where $G(s)$ is the transfer function matrix corresponding to

$$P(s) = \begin{pmatrix} sI_n - A & B \\ -C & 0 \end{pmatrix}$$

5 Let $G(s)$ be a proper $m \times l$ rational transfer function matrix, and let $d(s)$ with degree p be the least common denominator of its elements. Write $G(s) = N(s)/d(s)$, where $N(s)$ is a polynomial matrix, and let V_5 be the space spanned by the columns of

$$(N(s), sN(s) \bmod d(s), \ldots, s^{p-1}N(s) \bmod d(s))$$

Show that V_5 has dimension $\nu(G)$. [Use the results of Problems 1–4.]

6 Let $N(s)$ in Problem 5 be written

$$N(s) = N_0 + N_1 s + \cdots + N_{p-1}s^{p-1}$$

Show that the space V_6 spanned by the columns of

$$\begin{pmatrix} N_0 \\ N_1 \\ \vdots \\ N_{p-1} \end{pmatrix}$$

is isomorphic to V_5. Hence obtain Algorithm 5.2. [Note that $\nu(G^{\mathrm{T}}) = \nu(G)$. This was the original derivation.]

7 From equation (5.14) show that if $G(s) = N(s)/d(s)$ is proper, the polynomial part of $(G_k s^{-1} + G_{k+1}s^{-2} + \cdots)\, d(s)$ is $s^{k-1}N(s) \bmod d(s)$, where the notation is that of Problem 6. Hence show that the rank of the matrix H in (5.15) is the same as that of R in (5.7).

8 With the notation of Problem 5, let $d(s)$ have p distinct roots $\alpha_1, \alpha_2, \ldots, \alpha_p$, and write

$$G(s) = \sum_{i=1}^{p} \frac{K_i}{s - \alpha_i}$$

Show that $\nu(G)$ is $\sum_i$ rank K_i. [Use Theorem 6.4.]

9 By four different methods find $\nu(G)$ for

$$G(s) = \begin{pmatrix} \dfrac{s+1}{(s+2)(s+3)} & \dfrac{1}{s+1} \\ \dfrac{s+2}{(s+3)(s+4)} & \dfrac{s+4}{(s+1)(s+2)} \end{pmatrix}$$

10 Investigate the decoupling zeros of a direct analogue representation of $G(s)$ in Problem 9 and give the dimensions of the corresponding subspaces of the state space.

11 Let $G = T^{-1}U = T_1^{-1}U_1$ where T, U are relatively (left) prime polynomial matrices, and so are T_1, U_1. Show that there is a unimodular matrix M such that $T_1 = MT$, $U_1 = MU$. [Put $Q = T_1T^{-1}$, so that $T_1 = QT$, $U_1 = QU$. Bring Q to McMillan form, $LQR = E\Psi^{-1}$, and conclude from $\Psi LT_1 = ER^{-1}T$, $\Psi LU_1 = ER^{-1}U$ that $\Psi = E = I$; whence Q is unimodular.]

4 Linear constant discrete-time systems

In Chapters 2 and 3 the systems considered were governed by ordinary differential equations. We now turn to systems governed by difference equations. The theory of these runs so closely parallel to the theory already given that most of the detail may be omitted.

1 Difference equations

Let a dynamical system evolve discretely in time according to the equation

$$x_{k+1} = Ax_k + Bu_k \tag{1.1}$$

$$y_k = Cx_k + Du_k \tag{1.2}$$

Here x, u, y are vectors, respectively of dimension n, l, m; and A, B, C, D are real matrices. The index k takes on the values $0, 1, 2, \ldots$, and these are identified with times $t = 0, T, 2T, \ldots$.

Subject to a bound

$$\|u_k\| < M\rho^k \tag{1.3}$$

u has a z-transform $\tilde{u}$ (Chapter 1, Section 9).

$$\tilde{u}(z) = \sum_{k=0}^{\infty} u_k z^{-k} \tag{1.4}$$

valid for $|z| > \rho$. Then equations (1.1) and (1.2) can be transformed to give, for $x_0 = 0$,

$$z\tilde{x} = A\tilde{x} + B\tilde{u} \tag{1.5}$$

$$\tilde{y} = C\tilde{x} + D\tilde{u} \tag{1.6}$$

In contrast to equation (1.27) of Chapter 2, there is never any need to generalize (1.6) by writing $\tilde{D}(z)$ for D. If $\tilde{y}$ depended on $z\tilde{u}$, this would mean that a future input affected the present output. As our systems are assumed to be causal, this possibility is excluded.

A transfer function matrix $\tilde{G}(z)$ can be defined by

$$\tilde{G}(z) = C(zI_n - A)^{-1}B + D \tag{1.7}$$

and a system matrix can be written down

$$\tilde{P}(z) = \begin{pmatrix} zI_n - A & B \\ -C & D \end{pmatrix} \tag{1.8}$$

Then, exactly as in Chapter 2,

$$\tilde{G}^{i_1, i_2, \ldots, i_k}_{j_1, j_2, \ldots, j_k} = \frac{\tilde{P}^{i_1, i_2, \ldots, i_k}_{j_1, j_2, \ldots, j_k})}{\tilde{P})} \tag{1.9}$$

Indeed it will be clear that all results which can be obtained algebraically for a system matrix $P(s)$ in state-space form have their exact analogue for the $\tilde{P}(z)$ in (1.8): this also will be said to have state-space form.

Polynomial system matrices $\tilde{P}(z)$ can be defined as before by

$$\tilde{P}(z) = \begin{pmatrix} \tilde{T}(z) & \tilde{U}(z) \\ -\tilde{V}(z) & \tilde{W}(z) \end{pmatrix} \tag{1.10}$$

and the dimension r of $\tilde{T}$ will be assumed to satisfy $r \geqslant n$, where the order n of the system is the degree of $|\tilde{T}(z)|$. Rational system matrices are similarly defined.

The transformations of system similarity, strict system equivalence, and system equivalence can be defined just as before. So can least order, relative primeness, decoupling zeros, and the rest. All the conditions for T, U to be relatively (left) prime or T, V to be relatively (right) prime can be applied without change.

It would be tedious and unnecessary to list all these and the other results which carry over from Chapters 2 and 3 to the present situation. The reader will easily convince himself that apart from a few results such as those in Exercises 6.6 and 7.3 of Chapter 2 (which anticipate Chapter 5) nothing in those two chapters depends in any way upon the fact that a system matrix arises from differential equations.

Example 1.1 Let x_k satisfy

$$x_{k+1} = Ax_k + Bu_k \tag{1.11}$$

and let q be any integer not less than the degree of the minimal polynomial of A. By Theorem 6.2(v) of Chapter 2, $zI_n - A$ and B are relatively (left) prime if and only if the following condition is satisfied. Given any polynomial n-vector $\tilde{c}(z)$ with elements of degree $q - 1$ or less, there exist a polynomial n-vector $\tilde{x}(z)$ with elements of degree $q - 2$ or less, and a polynomial l-vector $\tilde{u}(z)$ with elements of degree $q - 1$ or less, such that

$$(zI_n - A)\tilde{x}(z) + B\tilde{u}(z) = \tilde{c}(z) \tag{1.12}$$

Let any n-vector x_0 be given, and put $\tilde{c}(z) = Ax_0z^{q-1}$. Then if $zI_n - A$ and B are relatively prime, (1.12) has a solution which we write

$$\tilde{x}(z) = x_1z^{q-2} + x_2z^{q-3} + \cdots + x_{q-1} \tag{1.13}$$

$$\tilde{u}(z) = -u_0z^{q-1} - u_1z^{q-2} - \cdots - u_{q-1} \tag{1.14}$$

Dividing (1.12) by z^{q-1} we obtain

$$(zI_n - A)(x_1z^{-1} + x_2z^{-2} + \cdots + x_{q-1}z^{-q+1}) - B(u_0 + u_1z^{-1} + \cdots + u_{q-1}z^{-q+1}) = Ax_0 \tag{1.15}$$

and equating coefficients of $z^0, z^{-1}, z^{-2}, \ldots, z^{-q+1}$

$$\left.\begin{aligned} x_1 &= Ax_0 + Bu_0 \\ x_2 &= Ax_1 + Bu_1 \\ &\vdots \\ x_{q-1} &= Ax_{q-2} + Bu_{q-2} \\ 0 &= Ax_{q-1} + Bu_{q-1} \end{aligned}\right\} \tag{1.16}$$

That is to say, given any initial state x_0 at $t = 0$, there is a finite sequence of inputs $u_0, u_1, \ldots, u_{q-1}$ which carries the state to the origin at time $t = qT$. This is the analogous result to that in Exercise 6.6 of Chapter 2. Notice that both results express the same algebraic fact; it is only the interpretation which depends on the nature of the underlying system.

EXERCISE 1.1 In the system (1.1), (1.2), let $u_1 = u_2 = \cdots = 0$, but let $u_0, u_{-1}, \ldots, u_{-n}$ be not necessarily zero: this amounts to moving the time origin to $-n$, and we assume $x_{-n} = 0$. Extend the summation in (1.4) from $k = -n$, so that $\tilde{u}(z)$ is a polynomial vector in z. Show that if $zI_n - A$, C are relatively (right) prime, then $y_1 = y_2 = \cdots = y_n = 0$ if and only if $\tilde{x}(z) = (zI_n - A)^{-1}B\tilde{u}(z)$ is polynomial. [Note that

$$\begin{pmatrix} y_1 \\ y_2 \\ \vdots \\ y_n \end{pmatrix} = \begin{pmatrix} Cx_1 \\ Cx_2 \\ \vdots \\ Cx_n \end{pmatrix} = \begin{pmatrix} C \\ CA \\ \vdots \\ CA^{n-1} \end{pmatrix} x_1,$$

from which the result follows.]

2 Sampling systems

Suppose that a system is given,

$$\dot{x} = Ax + Bu \tag{2.1}$$

$$y = Cx \tag{2.2}$$

but that the output y is sampled at the instants $t = 0, T, 2T, \ldots$. Suppose, too, that u is restricted to be constant over each interval $kT < t \leqslant (k + 1)T$. This would correspond, for example, to the situation when a digital computer was used for control purposes with sampling interval T. The output y would be observed, a calculation would be carried out, and a control vector u computed. This would be applied over the interval $kT < t \leqslant (k + 1)T$.

If the control vector over the interval $kT < t \leqslant (k + 1)T$ is u_k, the response of the system is

$$x_{k+1} = e^{AT}x_k + \int_0^T e^{A(T-\tau)}Bu_k \, d\tau \tag{2.3}$$

where $x_k = x(kT)$, etc. This gives

$$x_{k+1} = A_1x_k + B_1u_k \tag{2.4}$$

where, with a change of variable in (2.3),

$$A_1 = e^{AT} \tag{2.5}$$

$$B_1 = \int_0^T e^{A\tau} \, d\tau \, B \tag{2.6}$$

and equation (2.2) gives

$$y_k = Cx_k \tag{2.7}$$

Equations (2.4) and (2.7) correspond to equations (1.1) and (1.2), and the properties of the sampled system can be deduced from A_1, B_1, and C. For example a system matrix in state-space form corresponding to (2.4) and (2.7) is

$$\tilde{P}(z) = \begin{pmatrix} zI_n - A_1 & B_1 \\ -C & 0 \end{pmatrix} \tag{2.8}$$

If this has no decoupling zeros, the system has least order.

Questions of obvious interest are the relation of the decoupling zeros of (2.8) to those of

$$P(s) = \begin{pmatrix} sI_n - A & B \\ -C & 0 \end{pmatrix} \tag{2.9}$$

and the relation of the poles of

$$\tilde{G}(z) = C(zI_n - A_1)^{-1}B_1 \tag{2.10}$$

to those of

$$G(s) = C(sI_n - A)^{-1}B \tag{2.11}$$

Theorem 2.1 Let a system matrix $P(s)$ as in (2.8) have

transfer function poles $\{\alpha_i\}$, a in number;
i.d. zeros $\{\beta_i\}$, b in number;
o.d. zeros $\{\gamma_i\}$, c in number;
i.o.d. zeros $\{\delta_i\}$, d in number;

and suppose $T > 0$ to be such that $\mathrm{Im}\,(\alpha_i - \alpha_j)T/2\pi$ is not a non-zero integer for any α_i, α_j having $\mathrm{Re}\,\alpha_i = \mathrm{Re}\,\alpha_j$. Then the corresponding sampled system matrix (2.8) has

transfer function poles $\{e^{\alpha_i T}\}$, a in number;
decoupling zeros $\{e^{\beta_i T}, e^{\gamma_i T}\} - \{e^{\delta_i T}\}$, $b + c - d$ in number.

Proof Notice first that if the eigenvalues of any square matrix Q are $\lambda_1, \lambda_2, \ldots, \lambda_k$, then the eigenvalues of e^{QT} are $e^{\lambda_1 T}, e^{\lambda_2 T}, \ldots, e^{\lambda_k T}$. For if $K^{-1}QK = J$, where J is the Jordan form of Q, then from the series defining the matrix exponential it follows that

$$K^{-1}e^{QT}K = e^{K^{-1}QKT} = e^{JT} \tag{2.12}$$

The diagonal elements of e^{JT}, however, are $e^{\lambda_1 T}, e^{\lambda_2 T}, \ldots, e^{\lambda_k T}$.

Find a transformation of system similarity, as in Section 7.1 of Chapter 2, such that

$$\begin{pmatrix} H^{-1} & 0 \\ 0 & I_m \end{pmatrix}\begin{pmatrix} sI_n - A & B \\ -C & 0 \end{pmatrix}\begin{pmatrix} H & 0 \\ 0 & I_l \end{pmatrix} = \left(\begin{array}{c|c} sI_n - A_0 & B_0 \\ \hline -C_0 & 0 \end{array}\right)$$

$$= \left(\begin{array}{cccc|c} sI - A_{11} & -A_{12} & 0 & 0 & 0 \\ 0 & sI - A_{22} & 0 & 0 & 0 \\ -A_{31} & -A_{32} & sI - A_{33} & -A_{34} & B_3 \\ 0 & -A_{42} & 0 & sI - A_{44} & B_4 \\ \hline 0 & -C_2 & 0 & -C_4 & 0 \end{array}\right) \tag{2.13}$$

It is easily verified by direct multiplication that when A_0 has the special form defined in (2.13), so also do $A_0^2, A_0^3, \ldots$. That is to say, zero blocks in A_0 remain zero in all positive integral powers of A_0.

Now from (2.5) and (2.13)

$$H^{-1}A_1H = e^{H^{-1}AHT} = e^{A_0T} \tag{2.14}$$

and from the defining series for e^{A_0T} and what has just been said it follows that e^{A_0T} has the same form as A_0. It is also easy to verify that the diagonal (block) elements of e^{A_0T} are

$$e^{A_{11}T}, \qquad e^{A_{22}T}, \qquad e^{A_{33}T}, \qquad e^{A_{44}T} \tag{2.15}$$

Again, it is easy to confirm that A_0B_0, $A_0^2B_0, \ldots$ preserve the special form of B_0. Then from (2.6) and (2.13)

$$H^{-1}B_1 = \int_0^T H^{-1}e^{A\tau}H\, d\tau\; H^{-1}B \tag{2.16}$$

$$= \int_0^T e^{A_0\tau}\, d\tau\; B_0 \tag{2.17}$$

and it follows that $H^{-1}B_1$ has the same form as B_0. Consequently

$$\begin{pmatrix} H^{-1} & 0 \\ 0 & I_m \end{pmatrix}\begin{pmatrix} sI_n - A_1 & B_1 \\ -C & 0 \end{pmatrix}\begin{pmatrix} H & 0 \\ 0 & I_l \end{pmatrix} \tag{2.18}$$

has the same special form as in (2.13).

From (2.17) and the special form of A_0 and B_0, it will be seen that B_4 in (2.13) is replaced in (2.18) by

$$\int_0^T e^{A_{44}}\, d\tau\; B_4 = QB_4 \tag{2.19}$$

where Q commutes with A_{44}. Also A_{44} in (2.13) is replaced in (2.18) by $e^{A_{44}T}$. It follows that the transfer function poles in (2.18) will be $\{e^{\alpha_i T}\}$ if and only if the system matrix

$$\begin{pmatrix} sI_a - e^{A_{44}T} & QB_4 \\ -C & 0 \end{pmatrix} \tag{2.20}$$

has no decoupling zeros.

Let $H_1^{-1}A_{44}H_1 = J$ be in Jordan form. Then

$$|Q| = \left| \int_0^T e^{J\tau}\, d\tau \right| \tag{2.21}$$

and each element on the principal diagonal of the integral is T, if the corresponding $\alpha_i = 0$; or $e^{\alpha_i T} - 1$ if $\alpha_i \neq 0$. Because $T > 0$ it follows that $|Q| \neq 0$, and because Q commutes with A_{44} and so with $e^{A_{44}T}$, the i.d. zeros of (2.20) are the zeros of the Smith form of

$$H_1^{-1}Q^{-1}(sI_a - e^{A_{44}T} \quad QB_4)\begin{pmatrix} QH_1 & 0 \\ 0 & I_l \end{pmatrix} \tag{2.22}$$

$$= (sI_a - e^{JT} \quad H_1^{-1}B_4) \tag{2.23}$$

Let J_i be a $p \times p$ Jordan block in J corresponding to an eigenvalue α_i. In $sI_a - e^{JT}$ there is a corresponding block

$$\begin{pmatrix} s - e^{\alpha_i T} & -Te^{\alpha_i T} & \dots & -T^{p-1}e^{\alpha_i T}/(p-1)! \\ 0 & s - e^{\alpha_i T} & \dots & -T^{p-2}e^{\alpha_i T}/(p-2)! \\ \vdots & \vdots & & \vdots \\ 0 & 0 & \dots & s - e^{\alpha_i T} \end{pmatrix} \tag{2.24}$$

It is easy to see that the rank of (2.23) when $s = e^{\alpha_i T}$ will be a, if the rows of $H_1^{-1}B$ corresponding to the last rows of all Jordan blocks having eigenvalue $e^{\alpha_i T}$ are linearly independent. By the assumption on T, $e^{\alpha_i T} = e^{\alpha_j T}$ if and only if $\alpha_i = \alpha_j$. Consequently the condition for (2.23) to have rank a when $s = e^{\alpha_i T}$ is the same as the condition for

$$(sI_a - J \quad H_1^{-1}B_4) = H_1^{-1}(sI_a - A_{44} \quad B_4)\begin{pmatrix} H_1 & 0 \\ 0 & I_l \end{pmatrix} \tag{2.25}$$

to have rank a when $s = \alpha_i$: by the construction of (2.13) this condition is fulfilled. Consequently (2.20) has no i.d. zeros, and in the same way it has no o.d. zeros. It follows that (2.18) has as its transfer function poles $\{e^{\alpha_i T}\}$.

It is also clear from what has been said about the form of (2.18) that $\{e^{\beta_i T}\}$ are i.d. zeros, and $\{e^{\gamma_i T}\}$ are o.d. zeros. If some $\beta_k - \beta_j = 0 + i2\pi p$, $p = \pm 1, \pm 2, \dots$, it may happen that an i.d. zero in (2.13) which is not an i.o.d. zero, gives rise to an i.o.d. zero in (2.18). Then the set of decoupling zeros of (2.18) remains unchanged and equal to $\{e^{\beta_i T}, e^{\gamma_i T}\} - \{e^{\delta_i T}\}$. The same is true if extra i.o.d. zeros arise from the o.d. zeros of (2.13). This completes the proof.

Corollary 1 If T satisfies the condition of the theorem, $P(s)$ has least order if and only if $\tilde{P}(z)$ has least order.

Corollary 2 If T satisfies the condition of the theorem, the minimal polynomial of $e^{A_{44}T}$ has the same degree as the minimal polynomial of A_{44} in (2.13).

Proof It follows from (2.24) and the condition on T that the elementary divisors of $e^{A_{44}T}$ are obtained from those of A_{44} by substituting $e^{\alpha_i T}$ for α_i.

The reason for putting $D = 0$ in (2.2) is that if D is not zero, y responds immediately to a change in u. Then to preserve correspondence with what is usually done in practice, it is necessary to sample at $t = 0+, T+, 2T+, \dots$ rather than at $t = 0, T, 2T, \dots$. This would have unnecessarily obscured the details of the discussion, though it introduces no real difficulty.

EXERCISE 2.1 Show that if (2.2) is replaced by

$$y = Cx + Du \tag{2.26}$$

and if y is sampled at $0+, T+, 2T+, \ldots$, then (2.7) becomes

$$y_k = Cx_k + Du_k \tag{2.27}$$

[Notice that x is continuous.]

EXERCISE 2.2 Show that $\nu(\tilde{G}) \leqslant \nu(G)$, where $\tilde{G}$ and G are as in (2.10) and (2.11).

EXERCISE 2.3 A system has

$$G(s) = \begin{pmatrix} \dfrac{1}{s+1} \\ \dfrac{1}{s(s+2)} \end{pmatrix} \tag{2.28}$$

and is sampled at intervals of 0·1 sec. Find $\tilde{G}(z)$ and find state-space system matrices $P(s)$, $\tilde{P}(z)$ of least order giving rise to $G(s)$, $\tilde{G}(z)$.

3 System identification

So far in this book it has been assumed that a system will be characterized by its system matrix or transfer function matrix. Attention has been concentrated largely on methods for going from one description to any other. However, an important practical problem is to find such a description for a system by means of physical measurements made upon it. The initial state of the system is often unknown, and usually the measurements are subject to errors. Sometimes all the inputs and outputs are available before the calculation begins (off-line identification), but sometimes the calculations must be carried out as the information is received (on-line identification). In the second case a gradually improving estimate of the system equations is sought.

These problems will be treated in another volume, but the following simplified treatment has interesting connections with our present subject. Suppose that a discrete-time system, or a sampled differential system, is governed by the least-order equations

$$x_{k+1} = Ax_k + Bu_k \tag{3.1}$$

$$y_k = Cx_k \tag{3.2}$$

At $k = 0$ the state is $x_0 = 0$, and the output sequence y_k is observed when the input is $u_0 = e_i$, $u_1 = u_2 = \cdots = 0$. Here e_i is the ith column of I_l. Let the output sequences corresponding to the inputs $e_1, e_2, \ldots, e_l$ be assembled into the

$m \times l$ matrices $G_1, G_2, \ldots$. In other words, if $\tilde{G}(z)$ is the transfer function matrix of the system, let

$$\tilde{G}(z)I_l = G_1 z^{-1} + G_2 z^{-2} + \cdots \tag{3.3}$$

From knowledge of the matrices $G_1, G_2, \ldots$ we wish to find the transfer function matrix $\tilde{G}(z)$. Alternatively (and this comes to the same thing in view of Section 5 of Chapter 3) we wish to find any system matrix which is system similar to

$$\tilde{P}(z) = \begin{pmatrix} zI_n - A & B \\ -C & 0 \end{pmatrix} \tag{3.4}$$

If the infinite sequence $G_1, G_2, \ldots$ were given in its entirety we could simply sum (3.3) to obtain $\tilde{G}(z)$. In the present context, however, it is assumed that only a finite set $G_1, G_2, \ldots, G_q$ is available, but that these matrices are known exactly. We also assume that the order n of the underlying system is known: the fact that it is usually not known (together with the fact that the observations are subject to error) is one of the most important practical difficulties.

As in Section 1.3 of Chapter 3, minimal indices are defined by the equation

$$(zI_n - A \quad B)\begin{pmatrix} \tilde{x}(z) \\ -\tilde{u}(z) \end{pmatrix} = 0 \tag{3.5}$$

Here $\tilde{x}(z)$, $\tilde{u}(z)$ are polynomial vectors, and the highest degree occurring in them (which clearly occurs in $\tilde{u}(z)$) is $\lambda \leqslant n$. Equation (3.5) implies that the input $\tilde{u}(z)$ produces an output $\tilde{y}(z) = C\tilde{x}(z)$ which is polynomial in z. We shall write

$$\tilde{u}(z) = u_0 + u_{-1}z + \cdots + u_{-\lambda}z^{\lambda} \tag{3.6}$$

and shall develop two preliminary theorems.

Theorem 3.1 Two polynomial solution vectors of (3.5), $\begin{pmatrix} \tilde{x}_1 \\ -\tilde{u}_1 \end{pmatrix}$ and $\begin{pmatrix} \tilde{x}_2 \\ -\tilde{u}_2 \end{pmatrix}$, are linearly independent if and only if $\tilde{u}_1$, $\tilde{u}_2$ are linearly independent.

Proof Let α, β be any two polynomials. Then by (3.5)

$$B(\alpha\tilde{u}_1 + \beta\tilde{u}_2) = (zI_n - A)(\alpha\tilde{x}_1 + \beta\tilde{x}_2) \tag{3.7}$$

and because $|zI_n - A| \not\equiv 0$, $\alpha\tilde{x}_1 + \beta\tilde{x}_2 \equiv 0$ if $\alpha\tilde{u}_1 + \beta\tilde{u}_2 \equiv 0$ (compare Theorem 4.1 of Chapter 1). Conversely, if $\tilde{u}_1$ and $\tilde{u}_2$ are linearly independent, then *a fortiori* so are $\begin{pmatrix} \tilde{x}_1 \\ -\tilde{u}_1 \end{pmatrix}$ and $\begin{pmatrix} \tilde{x}_2 \\ -\tilde{u}_2 \end{pmatrix}$.

Theorem 3.2 Let $\tilde{u}(z)$ be a polynomial vector in z of degree not exceeding n, and let $\tilde{x}(z)$, $\tilde{y}(z)$ be defined by

$$\tilde{x}(z) = (zI_n - A)^{-1}Bu(z) \tag{3.8}$$

$$\tilde{y}(z) = C\tilde{x}(z) \tag{3.9}$$

where $zI_n - A$, C are relatively (right) prime. Then

$$\begin{pmatrix} G_1 & G_2 & \dots & G_{n+1} \\ G_2 & G_3 & \dots & G_{n+2} \\ \vdots & \vdots & & \vdots \\ G_n & G_{n+1} & \dots & G_{2n} \end{pmatrix} \begin{pmatrix} u_0 \\ u_{-1} \\ \vdots \\ u_{-n} \end{pmatrix} = 0 \tag{3.10}$$

if and only if $\tilde{x}(z)$ is a polynomial vector in z. Here u_{-i} is the coefficient of z^i in $\tilde{u}(z)$, and G_i is the coefficient of z^{-i} in the Laurent series expansion of $\tilde{G}(z) = C(zI_n - A)^{-1}B$.

Proof If $\tilde{x}(z)$ is polynomial, then $\tilde{y}(z)$ is polynomial by (3.9). From (3.8) and (3.9),

$$\begin{aligned} \tilde{y}(z) &= \tilde{G}(z)\tilde{u}(z) \\ &= (G_1 z^{-1} + G_2 z^{-2} + \cdots)(u_0 + u_{-1}z + \cdots + u_{-n}z^n) \end{aligned} \tag{3.11}$$

On equation to zero the coefficients of $z^{-1}, z^{-2}, \dots, z^{-n}$ in $\tilde{y}(z)$ we obtain (3.10). Conversely, if (3.10) is true then the coefficients of $z^{-1}, z^{-2}, \dots z^{-n}$ in $\tilde{y}(z)$ are zero, and as in Exercise 1.1 it follows that $\tilde{x}(z)$ is polynomial.

By Theorem 3.1, the linear independence of solutions of (3.5) can be determined by inspection of $\tilde{u}$ alone. By Theorem 3.2, if $\tilde{u}$ is polynomial and (3.10) is true, then $\tilde{x}$ satisfying (3.5) is also polynomial. Hence we may determine the linearly independent solutions of (3.5) by looking for those solutions of (3.10) which correspond to linearly independent $\tilde{u}(z)$. We may clearly assume $u_0 \neq 0$ in (3.10), for if u_0 were zero $\tilde{u}(z)$ would be linearly dependent on a polynomial vector of lower degree. Also we must remember, for example, that

$$\begin{pmatrix} u_0 \\ u_{-1} \\ u_{-2} \\ \vdots \\ u_{-p} \\ 0 \\ 0 \end{pmatrix}, \begin{pmatrix} 0 \\ u_0 \\ u_{-1} \\ \vdots \\ u_{-p+1} \\ u_{-p} \\ 0 \end{pmatrix}, \begin{pmatrix} 0 \\ 0 \\ u_0 \\ \vdots \\ u_{-p+2} \\ u_{-p+1} \\ u_{-p} \end{pmatrix}$$

all correspond to linearly dependent polynomial vectors. Subject to these remarks, we know by Theorem 1.1 of Chapter 3 that there are exactly l solutions of (3.10) corresponding to linearly independent polynomial vectors $\tilde{u}_1(z)$, $\tilde{u}_2(z), \dots, \tilde{u}_l(z)$, and these vectors have degrees $\lambda_1, \lambda_2, \dots, \lambda_l$ with $\lambda_1 + \lambda_2 + \cdots + \lambda_l = n$.

Now assemble the vectors so obtained into an $l \times l$ matrix $\tilde{T}(z)$

$$\tilde{T}(z) = (\tilde{u}_1(z) \quad \tilde{u}_2(z) \ldots \tilde{u}_l(z)) \tag{3.12}$$

Since the vectors $\tilde{u}_i(z)$ are linearly independent, $|\tilde{T}(z)| \not\equiv 0$. Also

$$\tilde{G}(z)\tilde{T}(z) = \text{poly}\,\{(G_1 z^{-1} + G_2 z^{-2} + \cdots + G z_{\lambda l}{}^{-\lambda_l})T(z)\} \tag{3.13}$$

$$= \tilde{V}(z) \tag{3.14}$$

where 'poly' is an instruction to take the polynomial part (in z) of the following expression: the columns of $\tilde{V}(z)$ are in fact the outputs corresponding to $\tilde{u}_1(z), \tilde{u}_2(z), \ldots, \tilde{u}_l(z)$. We have thus obtained

$$\tilde{G}(z) = \tilde{V}(z)\tilde{T}^{-1}(z) \tag{3.15}$$

It is easy to see that $\tilde{V}(z)$ and $\tilde{T}(z)$ are relatively (right) prime, for the degree of $|\tilde{T}(z)|$ in (3.10) cannot exceed $\lambda_1 + \lambda_2 + \cdots + \lambda_l = n$. Consequently $\delta(|\tilde{T}|) = n$, and the system $\tilde{T}\xi = \tilde{u}$, $\tilde{y} = \tilde{V}\xi$ has least order. In this procedure we have used $G_1, G_2, \ldots, G_{2n}$.

Example 3.1 Suppose that $n = 1$, $G_1 = (1, 2)$, $G_2 = (2, 4)$. Then (3.10) is

$$(1 \quad 2 \quad 2 \quad 4)\begin{pmatrix} u_0 \\ u_{-1} \end{pmatrix} = 0 \tag{3.16}$$

which has three linearly independent real vector solutions, say

$$\begin{pmatrix} u_0 \\ u_{-1} \end{pmatrix} = \begin{pmatrix} 2 \\ -1 \\ 0 \\ 0 \end{pmatrix}, \begin{pmatrix} 0 \\ 0 \\ 2 \\ -1 \end{pmatrix}, \begin{pmatrix} 2 \\ 0 \\ -1 \\ 0 \end{pmatrix} \tag{3.17}$$

Of these, the first and second correspond to the linearly dependent polynomial vectors $\begin{pmatrix} 2 \\ -1 \end{pmatrix}$ and $\begin{pmatrix} 2z \\ -z \end{pmatrix}$. We therefore obtain two linearly independent polynomial vectors $\tilde{u}_1(z) = \begin{pmatrix} 2 \\ -1 \end{pmatrix}$ with $\lambda_1 = 0$, and $\tilde{u}_2(z) = \begin{pmatrix} 2 - z \\ 0 \end{pmatrix}$ with $\lambda_2 = 1$. Then

$$\tilde{T}(z) = \begin{pmatrix} 2 & 2 - z \\ -1 & 0 \end{pmatrix} \tag{3.18}$$

and

$$\tilde{V}(z) = \text{poly}\left\{(z^{-1} \quad 2z^{-1})\begin{pmatrix} 2 & 2 - z \\ -1 & 0 \end{pmatrix}\right\} \tag{3.19}$$

$$= (0 \quad -1) \tag{3.20}$$

$$\tilde{G}(z) = (0 \quad -1)\begin{pmatrix} 2 & 2-z \\ -1 & 0 \end{pmatrix}^{-1} \tag{3.21}$$

$$= \begin{pmatrix} \frac{1}{z-2} & \frac{2}{z-2} \end{pmatrix} \tag{3.22}$$

EXERCISE 3.1 If an nth order system has one input and one output, show that equation (3.10) is satisfied by $u_{-n} = 1, u_{-n+1} = a_{n-1}, \ldots, u_0 = a_0$, where the characteristic polynomial of the system is $s^n + a_{n-1}s^{n-1} + \cdots + a_0$.

EXERCISE 3.2 Show that $\lambda_l = n$ if and only if B has rank 1.

EXERCISE 3.3 Show that by column interchanges (the same for both), $\tilde{V}$ and $\tilde{T}$ in (3.15) can be made to satisfy condition (ii) of Exercise 2.4, Chapter 3. [Use $\delta(|\tilde{T}|) = \lambda_1 + \lambda_2 + \cdots + \lambda_l$.]

EXERCISE 3.4 For each result given above for the λ_i, develop a parallel result for the minimal indices μ_i associated with the output. [See Corollary 1 of Theorem 1.1, Chapter 3. Start by restating each result for G^{T}.]

Having obtained $\tilde{G}(z)$ in the form $\tilde{V}(z)T^{-1}(z)$, a system matrix of least order for the system can be written immediately.

$$\tilde{P}(z) = \begin{pmatrix} I_{n-l} & 0 & 0 \\ 0 & T(z) & I_l \\ 0 & -V(z) & 0 \end{pmatrix} \tag{3.23}$$

in which I_{n-l} and the bordering zeros are to be omitted if $l \geqslant n$. The system matrix $\tilde{P}(z)$ can be brought to state-space form by strict system equivalence followed, if $l > n$, by a suitable reduction in size. If, however, the aim is to produce state-space equations, the following procedure is more direct.

The matrix H can be written, as in equation (5.18) of Chapter 3,

$$H = \begin{pmatrix} G_1 & G_2 & \ldots & G_n \\ G_2 & G_3 & \ldots & G_{n+1} \\ \vdots & \vdots & & \vdots \\ G_n & G_{n+1} & \ldots & G_{2n-1} \end{pmatrix} = \begin{pmatrix} C \\ CA \\ \vdots \\ CA^{n-1} \end{pmatrix} (B \quad AB \ldots A^{n-1}B) \tag{3.24}$$

and because the system (3.1), (3.2) has order n, each of the factors in (3.24) has rank n and so does their product. Consequently there exist constant non-singular matrices M, N, respectively $nm \times nm$, $nl \times nl$, such that

$$MHN = \begin{pmatrix} I_n & 0 \\ 0 & 0 \end{pmatrix} \tag{3.25}$$

Now let crosses represent blocks in which we have no present interest, and write

$$M\begin{pmatrix} C \\ CA \\ \vdots \\ CA^{n-1} \end{pmatrix} = \begin{pmatrix} L \\ \times \end{pmatrix} \tag{3.26}$$

$$(B \quad AB\ldots A^{n-1}B)N = (R \quad \times) \tag{3.27}$$

where L, R are $n \times n$ and (3.25) gives

$$LR = I_n \tag{3.28}$$

It follows from (3.26) to (3.28) and from (3.24) that

$$MH = \begin{pmatrix} R^{-1}B & \times \\ \times & \times \end{pmatrix} \tag{3.29}$$

$$MN = \begin{pmatrix} CR & \times \\ \times & \times \end{pmatrix} \tag{3.30}$$

Now define H' by

$$H' = \begin{pmatrix} G_2 & G_3 & \ldots & G_{n+1} \\ G_3 & G_4 & \ldots & G_{n+2} \\ \vdots & \vdots & & \vdots \\ G_{n+2} & G_{n+2} & \ldots & G_{2n} \end{pmatrix} \tag{3.31}$$

which is readily seen to be

$$H' = \begin{pmatrix} C \\ CA \\ \vdots \\ CA^{n-1} \end{pmatrix} A \quad (B \quad AB\ldots A^{n-1}B) \tag{3.32}$$

By our previous results we then have

$$MH'N = \begin{pmatrix} L \\ \times \end{pmatrix} A(R \quad \times) \tag{3.33}$$

$$= \begin{pmatrix} R^{-1}AR & \times \\ \times & \times \end{pmatrix} \tag{3.34}$$

Picking out the appropriate blocks from (3.29), (3.30), and (3.34) we have the required matrices

$$A_1 = R^{-1}AR, \quad B_1 = R^{-1}B, \quad C_1 = CR \tag{3.35}$$

The algorithm requires only the matrices H and H', from the first of which M and N are found by a standard procedure [compare Chapter 1, Section 1]. It therefore uses $G_1, G_2, \ldots, G_{2n}$, as before.

Example 3.2 For the system in Example 3.1, $H = (1, 2)$ and we may take

$$M = 1, \quad N = \begin{pmatrix} 1 & -2 \\ 0 & 1 \end{pmatrix} \tag{3.36}$$

Then

$$MH = (1 \quad 2) \tag{3.37}$$

$$HN = (1 \quad \times) \tag{3.38}$$

$$MH'N = 1(2 \quad 4)\begin{pmatrix} 1 & -2 \\ 0 & 1 \end{pmatrix} \tag{3.39}$$

$$= (2 \quad \times) \tag{3.40}$$

giving

$$A_1 = 2, \qquad B_1 = (1 \quad 2), \qquad C_1 = 1 \tag{3.41}$$

From this follows

$$\tilde{G}(z) = \left(\frac{1}{z-2} \quad \frac{2}{z-2}\right) \tag{3.42}$$

which agrees with (3.22).

Exercise 3.5 Show that n in (3.24) and (3.31) can be replaced by any $r > n$. Show that the resulting system will still have least order.

Exercise 3.6 Given

$$G(s) = \begin{pmatrix} \dfrac{1}{(s+1)^2} & \dfrac{1}{(s+1)(s+2)} \\ \dfrac{1}{(s+1)(s+2)} & \dfrac{s+3}{(s+2)^2} \end{pmatrix} \tag{3.43}$$

express the elements of $G(s)$ by their Laurent series and derive a state-space representation of least order from (3.35). Compare Example 5.3 in Chapter 3. [Find an upper bound for n and use the result of Exercise 3.3.]

Problems

1 Find the least order $\nu(G)$ of a system which can give rise to the transfer function matrix

$$\tilde{G}(z) = \begin{pmatrix} \dfrac{1}{z+1} & \dfrac{-z}{z+1} \\ \dfrac{1}{z+1} & \dfrac{1}{z+1} \end{pmatrix}$$

2 Give an example of a system matrix of least order giving rise to the $\tilde{G}(z)$ in Problem 1. Is the corresponding system stable? Is it asymptotically stable?

3 Euler's method for the solution of

$$\dot{x} = Ax + Bu$$

is to solve the associated difference equation, for some $h > 0$,

$$x_{k+1} = (I_n + hA)x_k + hBu_k$$

If $sI_n - A$, B are relatively (left) prime, show that $zI_n - I_n - hA$, hB are relatively (left) prime.

4 If in Problem 3, $\dot{x} = Ax$ is asymptotically stable, give a condition on h which ensures that $x_{k+1} = (I_n + hA)x_k$ is asymptotically stable.

5 By means of an example show that the condition on T in Theorem 2.1, though sufficient, is not necessary.

6 Let $\tilde{v}(z)$ be a polynomial l-vector, and let $\tilde{u}_1(z), \tilde{u}_2(z), \ldots, \tilde{u}_l(z)$ be the vectors occurring in (3.12). Show that $\tilde{G}(z)\tilde{v}(z)$ is a polynomial vector if and only if $\tilde{v}(z) = \sum_{i=1}^{l} \alpha_i(z)\tilde{u}_i(z)$, where the $\alpha_i(z)$ are polynomials. [This result has the following interpretation. As in Kalman, Falb, and Arbib (1969), regard the system as a mapping $f: \Omega \to \Gamma$. Here Ω is a finite free module over the ring of polynomials in z and its elements may be identified with the z-transforms of finite input sequences terminating at time 0. Also Γ is a module over the same ring: each element of Γ may be regarded as the z-transform (when this exists) of that part of an output sequence which occurs after time 0. The result is then equivalent to saying that the kernel of the mapping f is the submodule generated by $\tilde{u}_1, \tilde{u}_2, \ldots, \tilde{u}_l$.]

7 Let $G(z)$ be expressed in any way as $\tilde{V}_1(z)\tilde{T}_1^{-1}(z)$, where the polynomial matrices $\tilde{T}_1$, $\tilde{V}_1$ are not necessarily relatively (right) prime. Show that each column of $\tilde{T}_1$ can be expressed as a sum $\sum_{i=1}^{l} \alpha_i(z)\tilde{u}_i(z)$, where the $\alpha_i(z)$ are polynomials and the $\tilde{u}_i(z)$ are the polynomial vectors occurring in (3.12). Hence show that $\tilde{T}_1(z) = \tilde{T}(z)M(z)$, where $M(z)$ is a polynomial matrix and $\tilde{T}(z)$ is defined by (3.12). Also show that if $\tilde{T}_1$, $\tilde{V}_1$ are relatively (right) prime then $M(z)$ is unimodular. [Use the result of Problem 6. Compare with Problem 11, Chapter 3, and Exercise 3.5 of this chapter].

5 Stability and control

Attention will now be devoted to dynamical systems which are subject to control. The systems will be linear and time-invariant, and may be governed by differential or difference equations. Whenever control is applied by means of feedback, the question of stability becomes important. We shall therefore consider the stability of multivariable systems with feedback.

1 Closed-loop system matrix

Let a system have the equations, after Laplace transformation,

$$T(s)\bar{\xi} = U(s)\bar{u} \tag{1.1}$$

$$\bar{y} = V(s)\bar{\xi} + W(s)\bar{u} \tag{1.2}$$

where T, U, V, W are polynomial matrices, $|T|$ has degree n, and $\bar{\xi}$, $\bar{u}$, $\bar{y}$ have dimension respectively $r(\geqslant n)$, l, m, with $l = m$. From these equations follows

$$\bar{y} = [V(s)T^{-1}(s)U(s) + W(s)]\bar{u} = G(s)\bar{u} \tag{1.3}$$

Suppose that $\bar{v}$ is an m-vector of reference inputs, and let feedback be applied according to the equation

$$\bar{u} = \bar{v} - F(s)\bar{y} \tag{1.4}$$

Here F is an $m \times m$ rational matrix: in applications F is often constant and diagonal. From (1.3) and (1.4), if $|I_m + FG| \not\equiv 0$,

$$\bar{y} = G[I_m + FG]^{-1}\bar{v} = [I_m + GF]^{-1}G\bar{v} = H\bar{v} \tag{1.5}$$

where H is the closed-loop transfer function, which is also a rational system matrix for the closed-loop system. The first result in (1.5) is obtained by substituting $F\bar{y} = FG\bar{u}$ in (1.4) and the second by substituting (1.4) in $\bar{y} = G\bar{u}$.

Equations (1.1), (1.2), and (1.4) can be written

$$\left(\begin{array}{ccc|c} T & U & 0 & 0 \\ -V & W & -I_m & 0 \\ 0 & I_m & F & -I_m \\ \hline -V & W & 0 & 0 \end{array}\right)\left(\begin{array}{c} \bar{\xi} \\ -\bar{u} \\ -\bar{y} \\ \hline -\bar{v} \end{array}\right) = \left(\begin{array}{c} 0 \\ 0 \\ 0 \\ \hline -\bar{y} \end{array}\right) \tag{1.6}$$

The first matrix P_c on the left-hand side of (1.6) is a rational system matrix for the closed-loop system provided (as we now assume) that

$$\begin{vmatrix} T & U & 0 \\ -V & W & -I_m \\ 0 & I_m & F \end{vmatrix} \tag{1.7}$$

is not identically zero. If $F(s)$ is a polynomial matrix, then P_c qualifies as a polynomial system matrix provided (as we further assume) that the degree of the determinant (1.7) does not exceed $r + 2m$. If this last condition is not initially satisfied, it is easy to expand T, U, V in a trivial way so that the condition is fulfilled.

EXERCISE 1.1 Show that $|I_m + FG| = |I_m + GF|$; and that $|I_m + FG| \not\equiv 0$ in (1.5) if and only if the determinant (1.7) is not identically zero. [Consider $\begin{vmatrix} I_m & G \\ -F & I_m \end{vmatrix}$. Note the initial assumption that $|T| \not\equiv 0$.]

EXERCISE 1.2 Show that if $T = sI_n - A$, $U = B$, $V = C$, $W = 0$, and F is independent of s, the system matrix in (1.6) can be brought by strict system equivalence to the form

$$\left(\begin{array}{cc|c} I_{2m} & 0 & 0 \\ 0 & sI_n - A + BFC & B \\ \hline 0 & -C & 0 \end{array}\right) \tag{1.8}$$

EXERCISE 1.3 Show that if $sI_n - A$ and B in Exercise 1.2 are relatively (left) prime, then so also are $sI_n - A + BFC$ and B. State the corresponding result for $sI_n - A$ and C.

Theorem 1.1 Let a system be given as in (1.1) and (1.2) and let feedback be applied according to (1.4), with $F(s)$ a polynomial matrix such that the determinant (1.7) is not identically zero. Let

$$P_0(s) = \begin{pmatrix} T & U \\ -V & W \end{pmatrix} \tag{1.9}$$

have i.d. zeros $\beta_1, \beta_2, \ldots, \beta_b$, o.d. zeros $\gamma_1, \gamma_2, \ldots, \gamma_c$, and i.o.d. zeros $\delta_1, \delta_2, \ldots, \delta_d$. Then the matrix P_c in (1.6) has

(i) i.d. zeros $\beta_1, \beta_2, \ldots, \beta_b$ and no others;
(ii) o.d. zeros $\gamma_1, \gamma_2, \ldots, \gamma_c$ and no others;
(iii) i.o.d. zeros $\delta_1, \delta_2, \ldots, \delta_d$ and no others.

Proof The matrix

$$\begin{pmatrix} T & U & 0 & 0 \\ -V & W & -I_m & 0 \\ 0 & I_m & F & -I_m \end{pmatrix} \tag{1.10}$$

can be brought by elementary column operations to the form

$$\begin{pmatrix} T & U & 0 & 0 \\ 0 & 0 & I_m & 0 \\ 0 & 0 & 0 & I_m \end{pmatrix} \tag{1.11}$$

from which (i) follows at once.

The matrix

$$\begin{pmatrix} T & U & 0 \\ -V & W & -I_m \\ 0 & I_m & F \\ -V & W & 0 \end{pmatrix} \tag{1.12}$$

can be brought by elementary row operations to the form

$$\begin{pmatrix} T & U & 0 \\ 0 & 0 & -I_m \\ 0 & I_m & 0 \\ -V & 0 & 0 \end{pmatrix} \tag{1.13}$$

from which (ii) follows in the same way as before. The elementary operations which take (1.12) into (1.13) do not involve the first r rows. Consequently if the i.d. zeros are eliminated from $(T\ U)$ in (1.10) to leave $(Q\ R)$, the matrix (1.13) will have T replaced by Q and U by R. The remaining o.d. zeros of P_c will then be the zeros of $|N(s)|$ where

$$\begin{pmatrix} Q \\ -V \end{pmatrix} = \begin{pmatrix} Q_1 \\ -V_1 \end{pmatrix} N \tag{1.14}$$

and Q_1, V_1 are relatively (right) prime. These are just $\{\gamma_i\} - \{\delta_i\}$, proving (iii).

Corollary If P_0 has least order, then so does P_c.

Theorem 1.2 Let the open-loop system matrix P_0 in (1.9) have least order, and let $F(s)$ be a polynomial feedback matrix such that the determinant (1.7) is not identically zero. Let the McMillan form of the $m \times m$ open-loop transfer function matrix G be $\operatorname{diag}\big(\varepsilon_i(s)/\psi_i(s)\big)$. Then the McMillan form of the closed-loop transfer function matrix H is $\operatorname{diag}\big(\varepsilon_i(s)/\psi_i'(s)\big)$, where the Smith form of the matrix

$$\begin{pmatrix} T & U & 0 \\ -V & W & -I_m \\ 0 & I_m & F \end{pmatrix} \quad \text{is} \quad \begin{pmatrix} I_{r+m} & 0 \\ 0 & \operatorname{diag}\big(\psi_m'(s), \psi_{m-1}'(s), \ldots, \psi_1'(s)\big) \end{pmatrix} \tag{1.15}$$

Proof Because P_0 has least order, it follows from the Corollary of Theorem 1.1 that P_c has least order. By Theorem 4.1 of Chapter 3, the numerator polynomials in the McMillan form of H are the invariant polynomials in positions $r + 2m + 1$, $r + 2m + 2, \ldots, r + 3m$ in the Smith form of P_c. By elementary row and column operations, P_c can be brought successively to the forms

$$\begin{pmatrix} T & U & 0 & 0 \\ 0 & 0 & -I_m & 0 \\ 0 & 0 & 0 & -I_m \\ -V & W & 0 & 0 \end{pmatrix}, \quad \begin{pmatrix} I_m & 0 & 0 & 0 \\ 0 & I_m & 0 & 0 \\ 0 & 0 & T & U \\ 0 & 0 & -V & W \end{pmatrix} \tag{1.16}$$

This shows that the numerator polynomials in the McMillan form of H are $\varepsilon_1, \varepsilon_2, \ldots, \varepsilon_m$. The remaining part of the theorem follows from Theorem 4.1, Chapter 3.

Corollary The set of zeros $\{\eta_i'\}$ of the determinant (1.7) is

$$\{\eta_i'\} = \{\beta_i, \gamma_i, \alpha_i'\} - \{\delta_i\} \tag{1.17}$$

where $\{\beta_i\}$ is the set of i.d. zeros of P_0, $\{\gamma_i\}$ is the set of o.d. zeros of P_0, $\{\delta_i\}$ is the set of i.o.d. zeros of P_0, and $\{\alpha_i'\}$ is the set of poles of the closed-loop transfer function H.

Proof Remove the decoupling zeros $\{\beta_i, \gamma_i\} - \{\delta_i\}$ from P_0: this affects only the first (block) row and first (block) column of P_c, and removes zeros $\{\beta_i, \gamma_i\} - \{\delta_i\}$ from the determinant (1.7). The remaining system has least order, whence the result.

Example 1.1 Let P_0 be

$$P_0(s) = \left(\begin{array}{ccc:c} s+1 & 0 & 0 & 0 \\ 0 & s+2 & 0 & 0 \\ 0 & 0 & s+3 & 1 \\ \hdashline 0 & -1 & -1 & 1 \end{array}\right) \tag{1.18}$$

and put $F = 1$. The determinant

$$\begin{vmatrix} s+1 & 0 & 0 & 0 & 0 \\ 0 & s+2 & 0 & 0 & 0 \\ 0 & 0 & s+3 & 1 & 0 \\ 0 & -1 & -1 & 1 & -1 \\ 0 & 0 & 0 & 1 & 1 \end{vmatrix} \tag{1.19}$$

is $(s+1)(s+2)(s+\frac{7}{2})$ [compare (1.15)], while P_0 has $\{\beta_i\} = \{-1, -2\}$, $\{\gamma_i\} = \{-1\}$, $\{\delta_i\} = \{-1\}$, whence (1.17) gives $\{\alpha_i'\} = \{-\frac{7}{2}\}$. The open-loop transfer function is $G(s) = (s+4)/(s+3)$, giving

$$H(s) = \frac{s+4}{s+3}\left(1 + \frac{s+4}{s+3}\right)^{-1} = \frac{s+4}{2s+7} \tag{1.20}$$

EXERCISE 1.4 Show that when H is defined by (1.5),

$$|H(s)| = \begin{vmatrix} T & U & 0 & 0 \\ -V & W & -I_m & 0 \\ 0 & I_m & F & -I_m \\ -V & W & 0 & 0 \end{vmatrix} \div \begin{vmatrix} T & U & 0 \\ -V & W & -I_m \\ 0 & I_m & F \end{vmatrix} \tag{1.21}$$

$$= \begin{vmatrix} T & U \\ -V & W \end{vmatrix} \div \begin{vmatrix} T & U & 0 \\ -V & W & -I_m \\ 0 & I_m & F \end{vmatrix} \tag{1.22}$$

$$= \frac{|VT^{-1}U + W|}{|I_m + F(VT^{-1}U + W)|} \tag{1.23}$$

$$= \frac{|G|}{|I_m + FG|} \tag{1.24}$$

[To obtain (1.21) use equation (2.29) of Chapter 2; for (1.23) use equation (1.25) of Chapter 1.]

EXERCISE 1.5 Show that

$$\frac{|G(s)|}{|H(s)|} = \begin{vmatrix} T & U & 0 \\ -V & W & -I_m \\ 0 & I_m & F \end{vmatrix} \div |T| = \frac{\Pi_i(s - \alpha_i')}{\Pi_i(s - \alpha_i)} \tag{1.25}$$

where α_i are the poles of G and α_i' are the poles of H. Notice that cancellation may occur in (1.25) even if P_0 (and therefore P_c) has least order. [Use (1.22) and (1.17).]

The relationship (1.5) between H and G is more simply expressed in terms of their inverses H^{-1} and G^{-1}, when these exist. Because we wish to write the elements of these inverse matrices we define

$$\hat{H} = H^{-1}, \qquad \hat{G} = G^{-1} \tag{1.26}$$

and similarly for other matrices which will occur. The elements of H^{-1} can then be written in the normal way,

$$\hat{H} = (\hat{h}_{ij}) \tag{1.27}$$

and similarly for $\hat{G}$. With this notation, (1.5) gives

$$\hat{H} = \hat{G} + F \tag{1.28}$$

In the matrix $G = \hat{G}^{-1}$ we have

$$g_{ii}^{-1} = \frac{|\hat{G}|}{\hat{G}_{ii}} \tag{1.29}$$

where $\hat{G}_{ii}$ is the cofactor of $\hat{g}_{ii}$ in $\hat{G}$. On expanding $|\hat{G}|$ in terms of its ith row, (1.29) gives

$$g_{ii}^{-1} = \hat{g}_{ii} + \hat{G}_{ii}^{-1} \sum_{\substack{j=1 \\ j \neq 1}}^{m} \hat{g}_{ij}\hat{G}_{ij} \tag{1.30}$$

Similarly an expansion of $|\hat{G}|$ in terms of its ith column gives

$$g_{ii}^{-1} = \hat{g}_{ii} + \hat{G}_{ii}^{-1} \sum_{\substack{j=1 \\ j \neq 1}}^{m} \hat{g}_{ji}\hat{G}_{ji} \tag{1.31}$$

Corresponding formulae hold for h_{ii}^{-1}.

EXERCISE 1.6 Show from (1.30) or (1.31) that if G is triangular, $g_{ii}^{-1} = \hat{g}_{ii}$.

It was assumed initially that the number l of inputs was equal to the number m of outputs, so that G was square. It was also assumed that feedback was applied according to (1.4) without prior compensation. Both of these assumptions are easily removed. Let $G(s)$ be $m \times l$, and let $K(s)$ be an $l \times m$ compensating matrix. Then

$$Q(s) = G(s)K(s) \tag{1.32}$$

is an $m \times m$ matrix, which can take the place of G in (1.5). A system matrix for Q can be constructed as in Section 5 of Chapter 3. Notice that we cannot have

$|Q(s)| \not\equiv 0$ unless $l \geqslant m$. If $|Q(s)| \not\equiv 0$, write $\hat{Q} = Q^{-1}$, when formulae analogous to (1.30) and (1.31) are available.

EXERCISE 1.7 Show that all the results of this section apply to systems governed by difference equations, if only z is substituted for s.

2 Controllability

In engineering practice, a system is called controllable if it is possible to achieve the specified aims of control, whatever these may be. By extension, the system is said to be more or less controllable according to the ease or difficulty of exerting control. These ideas of controllability have been embodied in a number of mathematical definitions, which however do not exhaust the possible meanings.

2.1 Pointwise state controllability to the origin

Let a system in state-space form have the equations

$$\dot{x} = Ax + Bu \tag{2.1}$$

The system will be called *pointwise state controllable to the origin*, or *controllable (p.s.t.o.)*, if, given any state $x(0) = c$, there exists a time $t_1 > 0$ and a control u defined on $[0, t_1]$ such that $x(t_1) = 0$. If (2.1) is replaced by

$$x_{k+1} = Ax_k + Bu_k \tag{2.2}$$

the system will be called controllable (p.s.t.o.) if, given any $x_0 = c$, there exist an integer $p > 0$ and a sequence $u_0, u_1, \ldots, u_{p-1}$ such that $x_p = 0$.

Theorem 2.1 The system (2.1) is controllable (p.s.t.o.) if and only if $sI_n - A$ and B are relatively (left) prime. The time t_1 may be chosen arbitrarily and u can be chosen from the class D_R^1.

Proof Choose T as in Theorem 2.1 of Chapter 4 and let u have the constant value u_k over the interval $kT < t \leqslant (k+1)T$, $k = 0, 1, \ldots$: then u belongs to D_R^1 on an open interval $J \subset [0, jT]$ if it is zero for $t \leqslant 0$ and bounded on $[0, (j+1)T]$. The states $x_k = x(kT)$ satisfy

$$x_{k+1} = A_1 x_k + B_1 u_k \tag{2.3}$$

where A_1 and B_1 are given by equations (2.5), (2.6) of Chapter 4. By Corollary 1 of Theorem 2.1, Chapter 4, $zI_n - A_1$, B_1 are relatively (left) prime if $sI_n - A$, B are relatively (left) prime. It was shown in Example 1.1 of Chapter 4 that if $zI_n - A_1$, B_1 are relatively (left) prime, then for arbitrary c there exists a sequence $u_0, u_1, \ldots, u_{q-1}$ which makes $x_q = x(qT) = 0$. Here q is any number not less than the degree of the minimal polynomial of A_1, which is the same (Corollary 2 of Theorem 2.1, Chapter 4) as the degree of the minimal polynomial of A. Given any $t_1 > 0$, q and T may be chosen, subject to the constraints on T, such that $qT = t_1$.

Conversely, if $sI_n - A$, B are not relatively (left) prime, there exists a non-singular matrix H, by Chapter 2, Section 7, such that

$$H^{-1}AH = \begin{pmatrix} A_{11} & 0 \\ A_{21} & A_{22} \end{pmatrix}, \qquad H^{-1}B = \begin{pmatrix} 0 \\ B_2 \end{pmatrix} \tag{2.4}$$

Then writing

$$x = H\begin{pmatrix} \xi_1 \\ \xi_2 \end{pmatrix} \tag{2.5}$$

where ξ_1 is conformable with A_{11}, equation (2.1) becomes

$$\dot{\xi}_1 = A_{11}\xi_1 \tag{2.6}$$

$$\dot{\xi}_2 = A_{21}\xi_1 + A_{22}\xi_2 + B_2u \tag{2.7}$$

Choosing

$$x(0) = H\begin{pmatrix} d \\ 0 \end{pmatrix} \tag{2.8}$$

where $d \neq 0$, *we obtain from* (2.5) and (2.6),

$$x(t) = H\begin{pmatrix} e^{A_{11}t}\, d \\ \xi_2(t) \end{pmatrix} \tag{2.9}$$

and because $e^{A_{11}t}$ is not singular for any t, $x(t) \neq 0$ for any $t > 0$.

Corollary The system (2.2) is controllable (p.s.t.o.) if $zI_n - A$, B are relatively (left) prime: then an arbitrary state x_0 can be brought to the origin at time qT, where q is any integer not less than the degree of the minimal polynomial of A. The matrices $zI_n - A$, B are relatively (left) prime if the system (2.2) is controllable (p.s.t.o.) and A is non-singular (as A will be if it arises by sampling from the system (2.1)).

Proof The first part has been established in the proof of the theorem. The second follows on bringing A and B to the forms shown in (2.4), noting that A_{11} is non-singular, and concluding that $A_{11}^k\, d \neq 0$ for any k.

Example 2.1 That controllability (p.s.t.o.) need not imply that $zI_n - A$, B are relatively (left) prime when A is singular is shown by the following example.

$$A = \begin{pmatrix} -2 & 1 \\ -8 & 4 \end{pmatrix}, \qquad B = \begin{pmatrix} 1 \\ 4 \end{pmatrix} \tag{2.10}$$

Let the initial condition be

$$x_0 = \begin{pmatrix} c_1 \\ c_2 \end{pmatrix} \tag{2.11}$$

and choose

$$u_0 = 2c_1 - c_2 \tag{2.12}$$

Then

$$x_1 = \begin{pmatrix} -2 & 1 \\ -8 & 4 \end{pmatrix}\begin{pmatrix} c_1 \\ c_2 \end{pmatrix} + \begin{pmatrix} 1 \\ 4 \end{pmatrix}(2c_1 - c_2) = 0 \tag{2.13}$$

whereas

$$(B \quad AB) = \begin{pmatrix} 1 & 2 \\ 4 & 8 \end{pmatrix} \tag{2.14}$$

has rank 1. That is, $zI_2 - A$, B are not relatively prime, yet the system is controllable (p.s.t.o.).

What has been called here controllability (p.s.t.o.) is usually called just *controllability*, or sometimes *complete controllability*. When the second term is used, controllability is taken to mean that some, but not all, states can be brought to the origin.

2.2 Pointwise state controllability from the origin

The system (2.1) will be called *pointwise state controllable from the origin*, or *controllable* (*p.s.f.o.*), if, given any state c, there exist a time $t_1 > 0$ and a control u defined on $[0, t_1]$ such that if $x(0) = 0$, then $x(t_1) = c$. The system (2.2) will be called controllable (p.s.f.o.) if given any state c, there exist an integer $p > 0$ and $u_0, u_1, \ldots, u_{p-1}$ such that if $x_0 = 0$, then $x_p = c$.

Theorem 2.2 The system (2.1) is controllable (p.s.f.o.) if and only if $sI_n - A$, B are relatively (left) prime. The time t_1 can be chosen arbitrarily, and u can be chosen from the class D_R^1.

Proof The proof is essentially similar to that in Theorem 2.1, but for the sake of variety it will be set out in a more direct fashion. Choose T as in Theorem 2.1, Chapter 4, obtain (2.3), and take the z-transform with $x_0 = 0$ to obtain

$$(zI_n - A_1)\tilde{x}(z) = B_1\tilde{u}(z) \tag{2.15}$$

Write

$$\tilde{x}(z) = x_1z^{-1} + x_2z^{-2} + \cdots + x_pz^{-p} \tag{2.16}$$

$$\tilde{u}(z) = -u_0 - u_1z^{-1} - \cdots - u_{p-1}z^{-p+1} \tag{2.17}$$

and look for a solution of (2.15) which makes $x_p = c$. On equating powers of z, (2.15) gives

$$\left.\begin{array}{l} x_1 + B_1u_0 = 0 \\ -A_1x_1 + x_2 + B_1u_1 = 0 \\ \quad - A_1x_2 + x_3 + B_1u_2 = 0 \\ \quad\quad \vdots \\ \quad\quad - A_1x_{p-2} + x_{p-1} + B_1u_{p-2} = 0 \\ \quad\quad\quad - A_1x_{p-1} + B_1u_{p-1} = -c \end{array}\right\} \tag{2.18}$$

which (with a change in sign of the u_i) are just the successive equations obtained from (2.3) for $k = 0, 1, \ldots, p-1$. Multiply the first of equations (2.18) by A_1 and add it to the second; then multiply the second by A_1 and add it to the third; and so on, obtaining

$$\left.\begin{array}{l} x_1 + B_1u_0 = 0 \\ \quad x_2 + B_1u_1 + A_1B_1u_0 = 0 \\ \quad\quad x_3 + B_1u_1 + A_1B_1u_1 + A_1^2B_1u_0 = 0 \\ \quad\quad \vdots \\ \quad\quad\quad B_1u_{p-1} + A_1B_1u_{p-2} + \cdots + A_1^{p-1}B_1u_0 = -c \end{array}\right\} \tag{2.19}$$

where the last equation can be written.

$$(B_1 \quad A_1B_1 \quad \ldots \quad A_1^{p-1}B_1)\begin{pmatrix} u_{p-1} \\ u_{p-2} \\ \vdots \\ u_0 \end{pmatrix} = -c \tag{2.20}$$

By Corollaries 1 and 2 of Theorem 2.1, Chapter 4, A_1 and A have their minimal polynomials of the same degree α, and if $sI_n - A$, B are relatively (left) prime, then so are $zI_n - A_1$, B_1. Then (2.20) will have a solution whenever $p \geqslant \alpha$. Moreover, p and T can be chosen so that $pT = t_1$, for any given t_1.

Conversely, if $sI_n - A$, B are not relatively (left) prime, there exists a transformation (2.4), and so (2.6) gives $\xi_1(t) = 0$ and

$$x(t) = H\begin{pmatrix} 0 \\ \xi_2(t) \end{pmatrix} \tag{2.21}$$

Hence if $x(0) = 0$, it is impossible to have

$$x(t) = H\begin{pmatrix} d \\ 0 \end{pmatrix} \tag{2.22}$$

for any $d \neq 0$ and any $t > 0$.

Corollary The system (2.2) is controllable (p.s.f.o.) if and only if $zI_n - A$, B are relatively (left) prime; then there exists a sequence $u_0, u_1, \ldots, u_{q-1}$ which carries the state from 0 at $t = 0$ to any given c at $t = qT$. Here q is any number not less than the degree of the minimal polynomial of A.

Proof Sufficiency has been established in the proof of the theorem. If $zI_n - A$, B are not relatively (left) prime, A and B may be brought to the form given in (2.4). Then

$$x_k = H\begin{pmatrix} A_{11}^k 0 \\ \xi_2^{(kT)} \end{pmatrix} \neq H\begin{pmatrix} d \\ 0 \end{pmatrix} \tag{2.23}$$

for any $d \neq 0$ and any k.

What has been called here controllability (p.s.f.o.) is usually called *complete reachability* (*from the origin*).

2.3 Pointwise state controllability

The system (2.1) will be called *controllable* (*p.s.*) if, given any two states c_0, c_1, there exist a $t_1 > 0$ and a u defined on $[0, t_1]$ such that x is carried from c_0 at $t = 0$ to c_1 at $t = t_1$. System (2.2) will be called controllable (p.s.) if, given any two states c_0, c_1, there exist an integer p and a sequence $u_0, u_1, \ldots, u_{p-1}$ such that x is carried from c_0 at $t = 0$ to c_1 at $t = pT$.

Theorem 2.3 Controllability (p.s.) implies controllability (p.s.t.o.) and controllability (p.s.f.o.). Controllability (p.s.f.o.) implies controllability (p.s.). Controllability (p.s.t.o.) implies controllability (p.s.) for (2.1), or for (2.2) if A is non-singular. A necessary and sufficient condition for controllability (p.s.) of (2.1) or (2.2) is that $sI_n - A$, B are relatively (left) prime.

Proof For (2.1) or (2.2), controllability (p.s.) implies controllability (p.s.t.o.) and controllability (p.s.f.o.) by definition. For (2.1) or (2.2), controllability (p.s.f.o.) implies controllability (p.s.t.o.) by Theorems 2.1, 2.2, and their corollaries. Then x can be taken from c_0 to 0, and subsequently, by the time-invariance of the systems, from 0 to c_1. For (2.1), and for (2.2) if A is non-singular, controllability (p.s.t.o.) implies controllability (p.s.f.o.) and the proof is completed as before.

Corollary 1 If in (2.1) $sI_n - A$ and B are relatively (left) prime, any state c_0 at $t = 0$ can be transferred to any c_1 at arbitrary t_1 by a $u \in D_R^1$.

Proof The state can be transferred by a $u \in D_R^1$ from c_0 at $t = 0$ to 0 at $t = \frac{1}{2}t_1$, and then by a $u \in D_R^1$ from 0 at $t = \frac{1}{2}t_1$ to c_1 at $t = t_1$.

Corollary 2 If in (2.2) $zI_n - A$ and B are relatively (left) prime, then any state c_0 at $t = 0$ can be carried into any state c_1 at $t = qT$. Here q can be any integer not less than the degree of the minimal polynomial of A.

Proof Proceeding as in (2.15) to (2.19) we obtain in place of (2.20),

$$(B \quad AB \quad \ldots \quad A^{p-1}B)\begin{pmatrix} u_{p-1} \\ u_{p-2} \\ \vdots \\ u_0 \end{pmatrix} = A^p c_0 - c_1 \tag{2.24}$$

which has a solution whenever p is not less than the degree of the minimal polynomial of A.

Example 2.2 In (2.1), let

$$A = \begin{pmatrix} -1 & 0 \\ 1 & -2 \end{pmatrix}, \qquad B = \begin{pmatrix} 1 \\ 0 \end{pmatrix} \tag{2.25}$$

so that $sI_2 - A$, B are relatively (left) prime. Then A_1 and B_1 as defined in equations (2.5), (2.6) of Chapter 4 are

$$A_1 = e^{AT} = \begin{pmatrix} 1 & 0 \\ 1 & 1 \end{pmatrix}\begin{pmatrix} e^{-T} & 0 \\ 0 & e^{-2T} \end{pmatrix}\begin{pmatrix} 1 & 0 \\ -1 & 1 \end{pmatrix} \tag{2.26}$$

$$= \begin{pmatrix} e^{-T} & 0 \\ e^{-T} - e^{-2T} & e^{-2T} \end{pmatrix} \tag{2.27}$$

$$B_1 = \int_0^T e^{At}B\,dt = \begin{pmatrix} 1 - e^{-T} \\ \frac{1}{2}(1 - e^{-T})^2 \end{pmatrix} \tag{2.28}$$

Now let

$$x(0) = \begin{pmatrix} 0 \\ 1 \end{pmatrix} = c \tag{2.29}$$

and find u_0, u_1 so that

$$x(2T) = A_1(A_1c + B_1u_0) + B_1u_1 = \begin{pmatrix} 0 \\ 1 \end{pmatrix} \tag{2.30}$$

By direct calculation,

$$u_0 = \frac{2(1 + e^{-2T})}{e^{-T}(1 - e^{-T})}, \qquad u_1 = \frac{-2(1 + e^{-2T})}{(1 - e^{-T})} \tag{2.31}$$

$$x(T) = \begin{pmatrix} \dfrac{2(1 + e^{-2T})}{e^{-T}} \\[2ex] \dfrac{1 - e^{-T} + e^{-2T}}{e^{-T}} \end{pmatrix}, \qquad x(2T) = \begin{pmatrix} 0 \\ 1 \end{pmatrix} \tag{2.32}$$

Theorem 2.4 If $sI_n - A$, B are not relatively (left) prime, then a state c_0 at $t = 0$ of (2.1) can be carried into a state c_1 at $t = t_1$ if both c_0 and c_1 lie in the subspace Ω spanned by the columns of

$$(B \quad AB \quad \ldots \quad A^{q-1}B) \tag{2.33}$$

where q is any integer not less than the degree of the minimal polynomial of A. This can be achieved by a $u \in D_R^1$, and t_1 may be chosen arbitrarily.

Proof As in Theorem 2.1 of Chapter 4, there is a non-singular matrix H such that

$$H^{-1}AH = \begin{pmatrix} A_{11} & 0 \\ A_{21} & A_{22} \end{pmatrix}, \qquad H^{-1}B = \begin{pmatrix} 0 \\ B_2 \end{pmatrix} \tag{2.34}$$

and also, for A_1, B_1 defined by equations (2.5), (2.6) of Chapter 4,

$$H^{-1}A_1H = \begin{pmatrix} A'_{11} & 0 \\ A'_{21} & A'_{22} \end{pmatrix}, \qquad H^{-1}B_1 = \begin{pmatrix} 0 \\ B'_2 \end{pmatrix} \tag{2.35}$$

The blocks in the partitioned matrices (2.35) have the same dimensions as the corresponding blocks in (2.34). Proceeding as in (2.15) to (2.19) we obtain in place of (2.20)

$$(B_1 \quad A_1B_1 \quad \ldots \quad A_1^{p-1}B_1)\begin{pmatrix} u_{p-1} \\ u_{p-2} \\ \vdots \\ u_0 \end{pmatrix} = A_1^p c_0 - c_1 \tag{2.36}$$

whence by (2.35)

$$\begin{pmatrix} 0 & 0 & \ldots & 0 \\ B'_2 & A'_{22}B'_2 & \ldots & (A'_{22})^{p-1}B'_2 \end{pmatrix}\begin{pmatrix} u_{p-1} \\ u_{p-2} \\ \vdots \\ u_0 \end{pmatrix} = H^{-1}A_1^pH.H^{-1}c_0 - H^{-1}c_1 \tag{2.37}$$

As in Theorem 2.1 of Chapter 4, the matrix

$$(B'_2 \quad A'_{22}B'_2 \quad \ldots \quad (A'_{22})^{p-1}B'_2) \tag{2.38}$$

has no rank defect if $p \geqslant q'$, where q' is the degree of the minimal polynomial of A'_{22}: this is the same as that of A_{22} and does not exceed that of A. It follows from this and the form of $H^{-1}A_1^pH$ that (2.37) has a solution if

$$c_0 = H\begin{pmatrix} 0 \\ c'_0 \end{pmatrix}, \qquad c_1 = H\begin{pmatrix} 0 \\ c'_1 \end{pmatrix} \tag{2.39}$$

where the partitioning conforms to that in the first matrix of (2.37). By (2.34) it follows that

$$\text{sp}\,(B \quad AB \quad \dots \quad A^{p-1}B) = H\begin{pmatrix}0\\ w\end{pmatrix} \tag{2.40}$$

where w has the dimension of c_0' and c_1', and can be arbitrary if $p \geqslant q'$ defined above. As in the proof of Theorem 2.2, pT can be chosen equal to any $t_1 > 0$. This proves the theorem.

Corollary If $zI_n - A$, B are not relatively (left) prime, then a state c_0 at $t = 0$ of (2.2) can be carried into a state c_1 at $t = qT$ if both c_0 and c_1 lie in the subspace Ω spanned by the columns of

$$(B \quad AB \quad \dots \quad A^{q-1}B) \tag{2.41}$$

Here q can be any integer not less than the degree of the minimal polynomial of A.

Proof This is included in the proof of the theorem.

For an obvious reason Ω is called the *controllable subspace*. This explains the terminology introduced in Section 7.1, Chapter 2. Any state in Ω can clearly be transferred to the origin, and is said to be *controllable* (*p.s.t.o.*). Similarly any state in Ω can be reached from the origin, and is said to be *controllable* (*p.s.f.o.*). A state in Ω is also said to be *controllable* (*p.s.*). The conditions $c_0 \in \Omega$, $c_1 \in \Omega$ in Theorem 2.4 and its corollary are sufficient but not necessary, as is shown by the following example.

Example 2.3 Let

$$\left.\begin{aligned}\dot{x}_1 &= 0\\ \dot{x}_2 &= u\end{aligned}\right\} \tag{2.42}$$

$$c_0 = \begin{pmatrix}1\\0\end{pmatrix}, \qquad c_1 = \begin{pmatrix}1\\1\end{pmatrix} \tag{2.43}$$

then u: $u(t) = 1$ transfers c_0 at $t = 0$ to c_1 at $t = 1$, yet neither c_0 nor c_1 is in Ω.

EXERCISE 2.1 Show that (with q as in Theorem 2.4) a state c_0 at $t = 0$ of (2.1) can be transferred to c_1 at $t = t_1$ if and only if

$$e^{At_1}c_0 - c_1 \in \text{sp}\,(B \quad AB \quad \dots \quad A^{q-1}B) \tag{2.44}$$

Hence show that if one of c_0, c_1 is in Ω, then so also must be the other. Therefore no state not in Ω is either controllable (p.s.t.o.) or controllable (p.s.f.o.). [From the Taylor series for e^{At_1} show that if $c_0 \in \Omega$, then $e^{At_1}c_0 \in \Omega$. As in the proof of Theorem 2.1 show that if $c_0 \notin \Omega$, then $e^{At_1}c_0 \notin \Omega$.]

EXERCISE 2.2 Show that a state c_0 at $t = 0$ of (2.2) can be transferred to a state c_1 at $t = pT$ if and only if

$$A^p c_0 - c_1 \in \mathrm{sp}\,(B \quad AB \quad \ldots \quad A^{p-1}B) \tag{2.45}$$

Hence show that no state not in Ω is controllable (p.s.f.o.). Show that we cannot conclude such states are not controllable (p.s.t.o.). Notice that the corollaries of Theorems 2.1, 2.2, 2.3, and 2.4 give a lower bound q for the number p of steps needed for transference from one state to another when the states may be chosen arbitrarily. The result of this example may give a $p < q$ for the particular c_0 and c_1 in question. [Work direct from (2.23).]

Systems which have been called here controllable (p.s.) have been called elsewhere *strongly connected.*

2.4 Modal controllability

The system (2.1) will be called *modally controllable*, or *controllable* (*m*.), if given any monic polynomial $\phi(s)$ of degree n with real coefficients there exists an $m \times n$ matrix C such that

$$|sI_n - A + BC| = \phi(s) \tag{2.46}$$

It will be shown in Section 4 that a necessary and sufficient condition for controllability (m.) is that $sI_n - A$, B should be relatively (left) prime. As this is a purely algebraic result it applies equally to (2.2).

What has been called here controllability (m.) has been called elsewhere *complete mode controllability.*

2.5 Functional controllability

In many industrial situations the aim of control is to make the output vector y of a plant take a certain form as a function of time. For example the regulator problem is to make $y_i(t) = v_i(t)$, where the v_i are constant, or slowly changing, desired outputs.

Let a system have the $(r + m) \times (r + m)$ polynomial system matrix

$$P(s) = \begin{pmatrix} T(s) & U(s) \\ -V(s) & W(s) \end{pmatrix} \tag{2.47}$$

so that the $m \times m$ transfer function matrix is

$$G(s) = V(s)T^{-1}(s)U(s) + W(s) \tag{2.48}$$

This system will be called *functionally controllable*, or *controllable* (*f.*) if it satisfies the following condition. Let the McMillan degree $\delta(G)$ of $G(s)$ be p. Then given any y which is zero for $t < 0$, has its pth derivative $D_p y \in D_R$, and satisfies $\| D_p y(t) \| < Me^{\alpha t}$ for some M, α, and all t, there exists a u such that with $x(0) = 0$, u generates y. Similarly the system having

$$\tilde{P}(z) = \begin{pmatrix} \tilde{T}(z) & \tilde{U}(z) \\ -\tilde{V}(z) & \tilde{W}(z) \end{pmatrix} \tag{2.49}$$

will be called controllable (f.) if given any sequence $0, 0, \ldots, 0, y_p, y_{p+1}, \ldots,$ where $p = \delta(\tilde{G})$, there is a sequence $u_0, u_1, \ldots$ which generates $0, 0, \ldots, 0,$ $y_p, y_{p+1}, \ldots$ when $x_0 = 0$.

Theorem 2.5 The system (2.47) is controllable (f.) if and only if one of the following equivalent conditions is fulfilled.

(i) $|G(s)| \not\equiv 0$
(ii) $|P(s)| \not\equiv 0$

Proof Conditions (i) and (ii) are equivalent because $|G| = |P| \div |T|$. Under the conditions on y, $D_p y$ has a Laplace transform $\bar{q}$, and so y has Laplace transform $\bar{y} = \bar{q}/s^p$. Suppose that (i) is true. Then G^{-1} exists, and can be written

$$G^{-1}(s) = Q_1(s) + Q_2(s) \tag{2.50}$$

where $Q_1(s) \to 0$ as $s \to \infty$ and $Q_2(s)$ is a polynomial matrix. No element of $Q_2(s)$ has degree higher than $p = \delta(G)$, because

$$p = \delta(G^{-1}) = \delta(Q_1) + \delta[Q_2(s^{-1})] \tag{2.51}$$

Consequently

$$\bar{u}(s) = G^{-1}(s)\bar{y}(s) = s^{-p}G^{-1}(s)\bar{q}(s) \tag{2.52}$$

$$= \sum_{i,j} \frac{K_{ij}\bar{q}(s)}{(s - \alpha_i)^j} + K\bar{q}(s) \tag{2.53}$$

where $s^{-p}G^{-1}$ has been put into partial fractions. From (2.53) $u(t)$ is obtained as a sum of terms of the form $\beta e^{\alpha_i t} \int_0^t \int_0^t \cdots \int_0^t e^{-\alpha_i t} q(t)\, dt$ with 0, 1, or more integrations, and it easily follows that u exists and is in D_R. Then the output generated by this u has Laplace transform $G(G^{-1}\bar{y})$: that is, u generates y.

Suppose that $|G(s)| \equiv 0$. Write G in the standard form of equation 2.9, Chapter 3,

$$G(s) = T_1^{-1}(s)[U(s) + T_1(s)D(s)] = T_1^{-1}(s)U_1(s) \tag{2.54}$$

so that $\bar{y}$ and $\bar{u}$ are related by

$$T_1(s)\bar{y}(s) = U_1(s)\bar{u}(s) \tag{2.55}$$

Because $|U_1(s)| \equiv 0$, there is (at least) one $\varepsilon_i(s) \equiv 0$ in the Smith form

$$\text{diag}\,(\varepsilon_i(s)) = L(s)U_1(s)R(s) \tag{2.56}$$

of U_1. It follows that one row (at least) in $L(s)U_1(s)$ is identically zero, say row i. Choose y so that row i in $L(s)T_1(s)\bar{y}(s)$ is non-zero, which is possible because $|L(s)T_1(s)| \not\equiv 0$. Then clearly the differential equation

$$L(D)T_1(D)y = L(D)U_1(D)u \tag{2.57}$$

cannot be satisfied by any u.

Corollary The system (2.49) is controllable (f.) if and only if one of the following equivalent conditions is fulfilled.

(i) $|\tilde{G}(z)| \not\equiv 0$
(ii) $|\tilde{P}(z)| \not\equiv 0$

Proof Suppose $|\tilde{G}| \not\equiv 0$. Because $\delta(\tilde{G}^{-1}) = p$, $\tilde{G}^{-1}\tilde{y}$ represents a sequence which is zero for $t < 0$, and so qualifies as an input u: it also generates $\tilde{y}$. The remainder of the proof follows the same lines as before.

EXERCISE 2.3 Show that if the initial assumption that $l = m$ is removed, a necessary and sufficient condition for controllability (f.) is: G contains an $m \times m$ minor which is not identically zero. Hence a necessary condition is $l \geqslant m$. [For sufficiency put $l - m$ elements of u to zero. For necessity proceed as in the proof of Theorem 2.5.]

Example 2.4 Let G be

$$G(s) = \begin{pmatrix} \dfrac{s+1}{s+2} & \dfrac{s+1}{s+2} \\ \dfrac{1}{s+3} & \dfrac{1}{s+3} \end{pmatrix} = \begin{pmatrix} s+2 & 0 \\ 0 & s+3 \end{pmatrix}^{-1} \begin{pmatrix} s+1 & s+1 \\ 1 & 1 \end{pmatrix} \tag{2.58}$$

Then

$$\begin{pmatrix} 1 & -(s+1) \\ 0 & 1 \end{pmatrix} \begin{pmatrix} s+1 & s+1 \\ 1 & 1 \end{pmatrix} = \begin{pmatrix} 0 & 0 \\ 1 & 1 \end{pmatrix} \tag{2.59}$$

and y and u are related by

$$\begin{pmatrix} D+2 & -(D+1)(D+2) \\ 0 & D+3 \end{pmatrix} y = \begin{pmatrix} 0 & 0 \\ 1 & 1 \end{pmatrix} u \tag{2.60}$$

so that if $U(t)$ is the unit step at $t=0$,

$$y\colon y(t) = \begin{pmatrix} U(t)e^{-2t} \\ 0 \end{pmatrix} \tag{2.61}$$

cannot be generated by any u.

The equations corresponding to (2.47) are

$$T(s)\bar{\xi}(s) = U(s)\bar{u}(s) \tag{2.62}$$

$$\bar{y}(s) = V(s)\bar{\xi}(s) + W(s)\bar{u}(s) \tag{2.63}$$

from which we obtain

$$\left(\begin{array}{cc:c} T(s) & U(s) & 0 \\ -V(s) & W(s) & -I_m \\ \hdashline 0 & I_m & 0 \end{array}\right)\left(\begin{array}{c} \bar{\xi}(s) \\ -\bar{u}(s) \\ \hdashline -\bar{y}(s) \end{array}\right) = \left(\begin{array}{c} 0 \\ 0 \\ \hdashline -\bar{u}(s) \end{array}\right) \tag{2.64}$$

If the system is controllable (f.), the first matrix in (2.64), partitioned as shown, is therefore a polynomial system matrix for the inverse system: it will be written $\hat{P}(s)$, but notice that $\hat{P} \neq P^{-1}$. It follows from (2.64) that if $\hat{G} = G^{-1}$,

$$\hat{g}_{ij} = \frac{\hat{P}^{i}_{j})}{\hat{P})} \tag{2.65}$$

$$= \frac{(-1)^{i+j}P^{1,2,\ldots,j-1,j+1,\ldots,m}_{1,2,\ldots,i-1,i+1,\ldots,m})}{|P|} \tag{2.66}$$

which may be compared with Chapter 2, Exercise 2.3.

Theorem 2.6 A necessary and sufficient condition for $\hat{P}$ to have least order is that P has least order.

Proof The matrix

$$\begin{pmatrix} T & U & 0 \\ -V & W & -I_m \end{pmatrix} \tag{2.67}$$

is equivalent to

$$\begin{pmatrix} T & U & 0 \\ 0 & 0 & -I_m \end{pmatrix} \tag{2.68}$$

which has rank $r + m$ for all s if and only if $(T\ U)$ has rank r for all s. By this and a similar result for the first two (block) columns in P, the theorem follows.

If the system (2.1) is controllable (p.s.) it follows that x may be made to take any assigned values for $t = t_1, t_2, t_3, \ldots$, and that these times can be as closely spaced as we desire. From this it might seem that, with $y = x$, the system must be controllable (f.). That this is not so can be seen by the following example.

Example 2.5 Let

$$A = \begin{pmatrix} -1 & 0 \\ 1 & -2 \end{pmatrix}, \qquad B = \begin{pmatrix} 1 \\ 0 \end{pmatrix} \tag{2.69}$$

giving controllability (p.s.). Set $C = I_2$, giving

$$G(s) = \begin{pmatrix} \dfrac{1}{s+1} \\ \dfrac{1}{(s+1)(s+2)} \end{pmatrix} \tag{2.70}$$

which is not controllable (f.). This is in fact self-evident: if we have $x(0) = 0$ and require $\bar{y}_1 = \bar{x}_1 = 0$ this defines $\bar{u} = 0$. Then $\bar{y}_2 = \bar{x}_2$ is also 0 and cannot be chosen at will. What happens if we attempt to make $x_2 = 1$ is illustrated by Example 2.2. As $T \to 0$ in (2.31),

$$x(T) \to \begin{pmatrix} 4 \\ 1 \end{pmatrix}, \qquad x(2T) = \begin{pmatrix} 0 \\ 1 \end{pmatrix} \tag{2.71}$$

Imagine that from $x(0) = 0$ at $t = 0$, the state has been brought to $x_1(2T) = 0$, $x_2(2T) = 1$, which is possible because the system is controllable (p.s.). Then applying the control given in Example 2.2 over subsequent intervals we can ensure that $x_1(t) = 0$, $x_2(t) = 1$ at $t = 2T, 4T, 6T, \ldots$, however small T may be. But at $t = 3T, 5T, \ldots$ we then have $x_1(t) = 4$. Notice also that although $x_2(t) = 1$ at $t = 2T, 3T, 4T, \ldots$, $x_2(t)$ does not remain equal to 1 for all intermediate values of t. These facts explain the term 'pointwise' in the description of controllability (p.s.).

Example 2.6 The previous example shows that controllability (p.s.) does not imply controllability (f.) of the state considered as output. Conversely, a system can be controllable (f.) without being controllable (p.s.) as is shown by the example

$$\left.\begin{aligned} \dot{x}_1 &= 0 \\ \dot{x}_2 &= u \\ y &= x_2 \end{aligned}\right\} \tag{2.72}$$

For systems governed by difference equations the distinction between controllability (p.s.) and controllability (f.) is particularly clear. The former ensures that $y_q, y_{2q}, \ldots$ can be assigned arbitrarily, but not (in general) the intervening values of y.

What has here been called controllability (f.) is elsewhere called *functional reproducibility*.

2.6 Controllability (I.)

It is well known that a 1×1 transfer function with a zero at s_0 in the open right half-plane offers difficulty in applying feedback control. Such a transfer function has the same gain for imaginary s as if the zero were at $-\bar{s}_0$ in the left half-plane, but has greater phase shift, and is therefore called *non-minimum phase*.

In discussing zeros in the right half-plane or left half-plane, a special case has to be made of zeros on the imaginary axis. In themselves these do not offer difficulties of the kind just mentioned, though a zero at the origin offers difficulties of a different kind (Exercise 2.5). Nevertheless, an arbitrarily small change in the system matrix may move an imaginary zero into the right half-plane. For this reason, and for uniformity with the treatment of stability, imaginary zeros will be treated as though they were in the right half-plane. Thus the complex plane will be divided into an open left half-plane and a closed right half-plane (Chapter 1, Section 8).

With an $m \times l$ transfer function matrix the situation is more complicated than for a 1×1 transfer function. It may happen that the transfer function g_{ij} of one path through the system has no right half-plane zero, but that when feedback is applied through some other path, the original path shows such a zero. Conversely, a right half-plane zero may disappear as feedback is applied through some other path.

Example 2.7 Let

$$G(s) = \begin{pmatrix} \dfrac{1}{s+1} & \dfrac{2}{s+3} \\ \dfrac{1}{s+1} & \dfrac{1}{s+1} \end{pmatrix} \tag{2.73}$$

Then $g_{11} = 1/(s+1)$ which offers no difficulty. On the other hand if feedback is applied according to the equation $u_2 = -k_2 y_2$, (1.5) gives

$$H(s) = \begin{pmatrix} \dfrac{s^2 + (4-k_2)s + 3 + k_2}{(s+1)(s+3)(s+1+k_2)} & \dfrac{2(s+1)}{(s+3)(s+1+k_2)} \\ \dfrac{1}{s+1+k_2} & \dfrac{1}{s+1+k_2} \end{pmatrix} \tag{2.74}$$

When k_2 is large, the zeros of h_{11} are approximately 1, $k_2 - 5$, and so are both in the right half-plane. As $k_2 \to \infty$ for fixed s, $h_{11} \to (1-s)/(s+1)(s+3)$.

The situation is difficult to analyse in general, and the limiting result which follows is therefore useful as a guide.

Theorem 2.7 Let feedback be applied from $\bar{y}_{i_1}$ to $\bar{u}_{j_1}$, $\bar{y}_{i_2}$ to $\bar{u}_{j_2}, \ldots, \bar{y}_{i_k}$ to $\bar{u}_{j_k}$. For some given $s = s_0$, let the feedback be such that $\bar{y}_{i_1}, \bar{y}_{i_2}, \ldots, \bar{y}_{i_k}$ are maintained equal to zero when an input $\bar{u}_{j_{k+1}}$ is applied. Then the transfer function seen between $\bar{u}_{j_{k+1}}$ and $\bar{y}_{i_{k+1}}$ at $s = s_0$ will be

$$\frac{G^{i_1, i_2, \ldots, i_{k+1}}_{j_1, j_2, \ldots, j_{k+1}}}{G^{i_1, i_2, \ldots, i_k}_{j_1, j_2, \ldots, j_k}} \tag{2.75}$$

where the minors of G are evaluated at $s = s_0$.

Proof For the purposes of the proof renumber the inputs and outputs so that u_{j_q} becomes u_q, y_{i_p} becomes y_p. Then for $s = s_0$, by assumption,

$$\bar{y}_p = g_{p,k+1}\bar{u}_{k+1} + \sum_{q=1}^{k} g_{pq}\bar{u}_q = 0, \quad p = 1, 2, \ldots, k \tag{2.76}$$

$$\bar{y}_{k+1} = g_{k+1,k+1}\bar{u}_{k+1} + \sum_{q=1}^{k} g_{k+1,q}\bar{u}_q \tag{2.77}$$

On solving (2.76) for $\bar{u}_q$, $q = 1, 2, \ldots, k$ by Cramer's rule,

$$\bar{u}_q = \frac{-\bar{u}_{k+1}G^{1,2,\ldots,k}_{1,2,\ldots,q-1,k+1,q+1,\ldots,k}}{G^{1,2,\ldots,k}_{1,2,\ldots,k}} \tag{2.78}$$

and inserting this in (2.77)

$$\bar{y}_{k+1} = \frac{\bar{u}_{k+1}\left\{g_{k+1,k+1}G^{1,2,\ldots,k}_{1,2,\ldots,k} - \sum_{q=1}^{k} g_{k+1,q}G^{1,2,\ldots,k}_{1,2,\ldots,q-1,k+1,q+1,\ldots,k}\right\}}{G^{1,2,\ldots,k}_{1,2,\ldots,k}} \tag{2.79}$$

$$= \frac{\bar{u}_{k+1}\left\{g_{k+1,k+1}G^{1,2,\ldots,k}_{1,2,\ldots,k} + \sum_{q=1}^{k}(-1)^{k+1+q}g_{k+1,q}G^{1,2,\ldots,k}_{1,2,\ldots,q-1,q+1,\ldots,k+1}\right\}}{G^{1,2,\ldots,k}_{1,2,\ldots,k}} \tag{2.80}$$

$$= \frac{G^{1,2,\ldots,k+1}_{1,2,\ldots,k+1}}{G^{1,2,\ldots,k}_{1,2,\ldots,k}}\,\bar{u}_{k+1} \tag{2.81}$$

On reverting to the original notation, this is (2.75).

Corollary Let G be $m \times m$, and by means of feedback from outputs $1, 2, \ldots, i-1, i+1, \ldots, m$ to inputs $1, 2, \ldots, i-1, i+1, \ldots, m$ respectively let these outputs be maintained equal to zero for $s = s_0$. Then the transfer function from input i to output i at $s = s_0$ is $\hat{g}_{ii}^{-1}(s_0)$, where $\hat{g}_{ii}$ is the iith element of $\hat{G} = G^{-1}$.

Proof This follows on reordering the indices in the numerator of (2.75).

Notice that in (2.75) the indices $i_1, i_2, \ldots, i_k$ or the indices $j_1, j_2, \ldots, j_k$ can be reordered in any way without changing the value of the quotient. That is to say, the k feedback loops may be arranged in any fashion so long as the outputs $\bar{y}_{i_1}, \bar{y}_{i_2}, \ldots, \bar{y}_{i_k}$ are maintained equal to zero. In practice we can usually make these outputs small, but not zero, and we rely on continuity to ensure that (2.75) is approximately correct.

Example 2.8 For G given by (2.73), equation (2.75) gives

$$\frac{G_{12}^{12}}{G_2^2} = \frac{1 - i\omega}{(i\omega + 1)(i\omega + 3)} \tag{2.82}$$

This agrees with h_{11} in (2.74) when $k_2 \to \infty$ for fixed $s = i\omega$.

Let G have all its poles in the open left half-plane. Suppose for simplicity that $l = m$ and that feedback is applied according to (1.4) with

$$F = \begin{pmatrix} I_k & 0 \\ 0 & 0 \end{pmatrix} \tag{2.83}$$

the inputs and outputs being renumbered to conform with (2.83). Suppose also that the feedback is so effective that

$$H(s) = G[I + FG]^{-1} \doteqdot \begin{pmatrix} I_k & 0 \\ H_{21} & H_{22} \end{pmatrix} \tag{2.84}$$

for all $|s| < \alpha$. Finally suppose that the numerator polynomials $\varepsilon_1, \varepsilon_2, \ldots, \varepsilon_k$ in the McMillan form of G have no zero in the closed right half-plane, but ε_{k+1} has such a zero at $s = s_1$, $|s_1| < \alpha$. By Theorem 1.2, every $(k + 1) \times (k + 1)$ minor of H has a zero at s_1. If H is assumed to have (exactly) the form given by (2.84) for all $|s| < \alpha$, it follows that every element of H_{22} has a zero at s_1. Like (2.75) this is a limiting result, useful because it gives some guidance in a complicated situation.

EXERCISE 2.4 Show that for G given by (2.73), ε_1 has no right half-plane zero but that ε_2 has such a zero. Confirm that every one of the four possible loops which can be set up gives rise, when its gain is high enough, to a right half-plane zero in the remaining path through the system.

On the other hand, if none of $\varepsilon_1, \varepsilon_2, \ldots, \varepsilon_m$ has a zero in the closed right half-plane, it is always possible by suitable compensation of the system to find m simultaneous paths with no right half-plane zeros. At worst this can be done by bringing G to triangular form.

In view of these results we shall call a system *controllable* (*l.*) if it is controllable (f.) and none of $\varepsilon_1, \varepsilon_2, \ldots, \varepsilon_m$ has a zero in the closed right half-plane.

EXERCISE 2.5 A system has $G(s) = s/(s + 1)^2$ and is therefore controllable (f.). Suppose that the desired output (from zero initial conditions) is

$$y(t) = (3t^2 - 2t^3)[U(t) - U(t - 1)] + U(t - 1) \tag{2.85}$$

which satisfies the conditions in Section 2.5. Show that for $t > 1$, u satisfies $\dot{u} = 1$, so that no u which is constant for $t > 1$ can give rise to the output in (2.85).

EXERCISE 2.6 Let the $m \times l$ transfer function matrix $G(s)$ of a system be written in the standard form (as in equation (2.54)) $G(s) = T^{-1}(s)U(s)$. Let the desired output for $t > t_1 > 0$ be $y(t) = c$, an arbitrary constant m-vector. Show that if the system is controllable (f.), the desired output for $t > t_1$ can be achieved with constant $u(t)$, $t > t_1$, if and only if $U(0)$ has rank m. Conclude that this will be true if the system is controllable (l.). [If $y(t) = c$ can be achieved with $u(t) = d$, a constant l-vector, then $T(D)c = U(D)d$, whence $T(0)c = U(0)d$ must have a solution d for all c.]

Example 2.9 It is not sufficient, in Exercise 2.6, to substitute (when $l = m$) the requirement $|G(0)| \neq 0$ for the condition $|U(0)| \neq 0$. For example,

$$G(s) = \begin{pmatrix} \dfrac{s}{(s+1)^2} & 0 \\ 0 & \dfrac{1}{s} \end{pmatrix} \tag{2.86}$$

has $|G(0)| = 1$, yet no constant u can achieve $y = \begin{pmatrix} 1 \\ 0 \end{pmatrix}$.

For systems governed by a difference equation, and with z-transfer function matrix $\tilde{G}(z)$, analogous considerations arise. Such a system will be called controllable (l.) if it is controllable (f.) and none of the numerator polynomials $\varepsilon_i(z)$ in the McMillan form of $\tilde{G}$ has a zero on or outside the unit circle.

2.7 Comment on terminology

There must always be reluctance to suggest changes in accepted terminology. In this area, however, the terminology is in some disarray and it seemed confusing to perpetuate it. The chief point which needs to be stressed is that controllability is an engineering term with a wide connotation. To restrict its meaning to one particular type of controllability seems wrong, and leads to confusion. For example, most industrial plants are controlled quite satisfactorily though they are not controllable (p.s.).

Much of the importance of controllability (p.s.t.o.) is that it implies for system

(2.1) that there are no i.d. zeros, and therefore there is no redundancy in the description of the system. For system (2.2), however, it is controllability (p.s.f.o.) which has this property. This is the natural criterion to use: only those states which can be reached from the origin contribute to the input–output behaviour with zero initial conditions. For system (2.1) controllability (p.s.t.o.) and controllability (p.s.f.o.) are equivalent, but in general one would have to use the latter ('reachability') as the criterion. It seems simpler to combine both types of controllability into controllability (p.s.) as was done here. The two types can be distinguished as (p.s.t.o.) and (p.s.f.o.) when required, but this distinction is seldom needed.

Another objection to current terminology is that although it is natural to say that a state is reachable, or an output is reproducible, it is not natural to apply these terms to a system. Finally, 'complete' seems to be redundant, in that if a system is (in usual terms) 'controllable' but not 'completely controllable', the qualification conveys only the negative information that some states cannot be brought to the origin. To be useful, it is necessary to say which states can be brought to the origin.

Whether or not the reader accepts the terms suggested, it is hoped that their use here has clarified the ideas involved.

3 Observability

Let a system be given with no input, and with output y:

$$\dot{x} = Ax \tag{3.1}$$

$$y = Cx \tag{3.2}$$

This system will be called *observable* if there exists a $t_1 > 0$ such that given y on the interval $[0, t_1]$ it is possible to deduce $x(0)$. If (3.1) (3.2) are replaced by

$$x_{k+1} = Ax_k \tag{3.3}$$

$$y_k = Cx_k \tag{3.4}$$

the system will be called observable if there exists a q such that given $y_0, y_1, \ldots, y_{q-1}$ it is possible to deduce x_0. If in either system there is an input u (so that Bu is added in (3.1) or Bu_k in (3.3)) then knowledge of u on the interval $[0, t_1]$, or of $u_0, u_1, \ldots, u_{q-1}$, as the case may be, allows the effect of u on y to be eliminated.

Theorem 3.1 The system (3.1), (3.2) or (3.3), (3.4) is observable if and only if $sI_n - A$, C are relatively (right) prime. For (3.1), (3.2), t_1 may be chosen arbitrarily. For (3.3), (3.4), q may be any number not less than the degree of the minimal polynomial of A.

Proof Equations (3.3), (3.4) give

$$\begin{pmatrix} y_0 \\ y_1 \\ \vdots \\ y_{q-1} \end{pmatrix} = \begin{pmatrix} C \\ CA \\ \vdots \\ CA^{q-1} \end{pmatrix} x_0 \tag{3.5}$$

If $sI_n - A$, C are relatively (right) prime, this equation can be solved for x_0 whenever q is not less than the degree of the minimal polynomial of A.

For (3.1), (3.2), choose T and q so that $qT = t_1$, q is as above, and T satisfies the conditions of Theorem 2.1, Chapter 4. Then the equations give

$$x_{k+1} = A_1 x_k \tag{3.6}$$

$$y_k = Cx_k \tag{3.7}$$

where $x_k = x(kT)$, $y_k = y(kT)$, $A_1 = e^{AT}$. Also $sI_n - A_1$, C are relatively (right) prime if $sI_n - A$, C are relatively (right) prime, while A, A_1 have their minimal polynomials of the same degree. Then $x_0 = x(0)$ can be obtained as in (3.5).

If $sI_n - A$, C in (3.1), (3.2) are not relatively (right) prime, then there exists H such that

$$\begin{pmatrix} H^{-1} & 0 \\ 0 & I_m \end{pmatrix}\begin{pmatrix} A \\ -C \end{pmatrix} H = \begin{pmatrix} A_{11} & A_{12} \\ 0 & A_{22} \\ \hline 0 & -C_2 \end{pmatrix} \tag{3.8}$$

Any $x(0)$ of the form

$$x(0) = H\begin{pmatrix} c \\ 0 \end{pmatrix} \tag{3.9}$$

with c conformable for $A_{11}c$ gives rise to $y = 0$, because

$$H^{-1}x(t) = H^{-1}e^{At}x(0) = \begin{pmatrix} e^{A_{11}t}c \\ 0 \end{pmatrix} \tag{3.10}$$

Therefore such an $x(0)$ cannot be determined from knowledge of y.

If $sI_n - A$, C in (3.3), (3.4) are not relatively (right) prime, there exists an H as above so that (3.8) is true. Any x_0 of the form (3.9) gives rise to

$$H^{-1}x_k = \begin{pmatrix} A_{11}^k c \\ 0 \end{pmatrix}, \qquad y_k = 0 \tag{3.11}$$

and so x_0 cannot be determined from y.

Notice that if $x(0)$ can be determined for (3.1), or x_0 for (3.3), then all later $x(t)$ or x_k can be obtained. It would be possible to define an alternative type of 'observability' in which $x(t_1)$ or x_q was to be determined rather than $x(0)$ or x_0, but for the reason just given this is not required. With such an alternative definition in (3.3), ability to determine x_q need not, if A is singular, imply that $sI_n - A$, C are relatively (right) prime. The situation would therefore be analogous to that found earlier for controllability (p.s.t.o.). This alternative type of 'observability' has also been called observability, but is usually distinguished as *constructibility* or *determinability*.

Example 3.1 In (3.3), (3.4), let

$$A = \begin{pmatrix} 1 & -1 \\ 1 & -1 \end{pmatrix}, \qquad C = (1 \quad -1) \tag{3.12}$$

Then for any x_0 whatever, $x_2 = x_3 = \cdots = 0$. So x_q may be determined for $q \geqslant 2$, yet it is easy to see that $sI_2 - A$, C are not relatively (right) prime.

Theorem 3.2 If $sI_n - A$, C in (3.1), (3.3) are not relatively (right) prime, then any state contained at $t = 0$ in the subspace Ω orthogonal to every row of the matrix

$$\begin{pmatrix} C \\ CA \\ \vdots \\ CA^{q-1} \end{pmatrix} \tag{3.13}$$

remains in Ω and gives zero output at all later times. Here q is any integer not less than the degree of the minimal polynomial of A.

Proof There exists an H such that

$$H^{-1}AH = \begin{pmatrix} A_{11} & A_{12} \\ 0 & A_{22} \end{pmatrix}, \qquad CH = (0 \quad C_2) \tag{3.14}$$

and $sI - A_{22}$, C_2 are relatively (right) prime. Then if

$$\begin{pmatrix} C \\ CA \\ \vdots \\ CA^{q-1} \end{pmatrix} x(0) = 0 \tag{3.15}$$

$$\begin{pmatrix} CH \\ CH.H^{-1}AH \\ \vdots \\ CH.H^{-1}A^{q-1}H \end{pmatrix} H^{-1}x(0) = 0 \tag{3.16}$$

so that if

$$x = H\begin{pmatrix} \xi^1 \\ \xi^2 \end{pmatrix} \tag{3.17}$$

with ξ^1 conformable for $A_{11}\xi^1$, (3.16) gives $\xi^2(0) = 0$, $\xi^1(0)$ arbitrary. Then for (3.1)

$$x(t) = e^{At}x(0) = H.e^{H^{-1}AHt}\begin{pmatrix} \xi^1(0) \\ \xi^2(0) \end{pmatrix} \tag{3.18}$$

$$= H\begin{pmatrix} e^{A_{11}t}\xi^1(0) \\ 0 \end{pmatrix} \tag{3.19}$$

if $\xi^2(0) = 0$. For (3.3),

$$x_k = A^k x_0 = H.(H^{-1}AH)^k\begin{pmatrix} \xi_0^1 \\ \xi_0^2 \end{pmatrix} \tag{3.20}$$

$$= H\begin{pmatrix} A_{11}^k \xi_0^1 \\ 0 \end{pmatrix} \tag{3.21}$$

if $\xi_0^2 = 0$. This completes the proof.

The subspace Ω is called the *unobservable subspace*, which explains the terminology introduced in Section 7.1, Chapter 2.

3.1 State observers

Let a system be given with

$$\dot{x} = Ax + Bu \tag{3.22}$$

$$y = Cx \tag{3.23}$$

where A, B, C are respectively $n \times n$, $n \times l$, $m \times n$, and rank C is assumed to be m: if rank $C < m$, some of the outputs are redundant and can be omitted. Suppose that for purposes of control we wish to know x. For example we might wish to apply optimal control in a form which assumed knowledge of x, or we might wish to allocate the poles of the closed-loop system as in Section 4. If our system equations had included noise in the measurements, the appropriate techniques would be statistical: for example the Kalman–Bucy filter might be applied. The problem which will be considered here is to deduce x from knowledge of u and y on the interval $[0, t]$ in the deterministic system (3.22), (3.23). We shall attempt to do this by finding a second dynamical system which, when driven by y and u, has a state which together with y gives an approximation to x. Such a system is called a *state observer*.

If in (3.23) C were square and non-singular, the problem would have the elementary solution

$$x = C^{-1}y \tag{3.24}$$

If on the other hand $C = 0$, the best we can do is to set up a replica of (3.22) and drive it with the same input,

$$\dot{\xi} = A\xi + Bu \tag{3.25}$$

Then

$$\frac{d}{dt}(\xi - x) = A(\xi - x) \tag{3.26}$$

and if A has all its eigenvalues in the open left half-plane, $\xi(t) \to x(t)$ as $t \to \infty$ regardless of the initial conditions for ξ and x. This will be the situation for the unobservable component of x if the system (3.22), (3.23) is not observable.

Let the matrices A_0, B_0, B_1, C_0, C_1 be respectively $(n - m) \times (n - m)$, $(n - m) \times l$, $(n - m) \times m$, $n \times (n - m)$, $n \times m$ and consider

$$\dot{\xi} = A_0\xi + B_0u + B_1y \tag{3.27}$$

$$\eta = C_0\xi + C_1y \tag{3.28}$$

If H is an $(n - m) \times n$ matrix, (3.22), (3.23), and (3.27) give

$$\dot{\xi} - H\dot{x} = A_0\xi - HAx + B_1Cx + (B_0 - HB)u \tag{3.29}$$

Suppose that H, A_0, B_0, and B_1 can be chosen so that

$$A_0H = HA - B_1C, \qquad B_0 = HB \tag{3.30}$$

Then (3.29) becomes

$$\frac{d}{dt}(\xi - Hx) = A_0(\xi - Hx) \tag{3.31}$$

and if all the eigenvalues of A_0 are in the open left half-plane,

$$\xi(t) \to Hx(t) \quad \text{as } t \to \infty \tag{3.32}$$

regardless of the initial conditions for ξ and x. Then if

$$(C_0 \quad C_1)\begin{pmatrix} H \\ C \end{pmatrix} = I_n \tag{3.33}$$

equations (3.28) and (3.32) show that

$$\eta(t) \to x(t) \quad \text{as } t \to \infty \tag{3.34}$$

The first of conditions (3.30) can be written

$$(A_0 \quad B_1)\begin{pmatrix} H \\ C \end{pmatrix} = HA \tag{3.35}$$

and with (3.33) given, (3.35) is true if and only if

$$(A_0 \quad B_1) = HA(C_0 \quad C_1) \tag{3.36}$$

All possible matrices A_0, B_0, B_1, C_0, C_1, H satisfying (3.30) and (3.33) can therefore be generated by the following procedure.

(i) Choose H so that $\begin{pmatrix} H \\ C \end{pmatrix}$ is non-singular.

(ii) Compute C_0, C_1 from

$$(C_0 \quad C_1) = \begin{pmatrix} H \\ C \end{pmatrix}^{-1} \tag{3.37}$$

(iii) Compute A_0, B_1 from (3.36).
(iv) Compute $B_0 = HB$.

On putting $m = n$ this procedure leads to

$$0 = H = A_0 = B_0 = B_1 = C_0, \qquad C_1 = C^{-1} \tag{3.38}$$

which agrees with (3.24). When $m = 0$, the result is

$$C = 0 = B_1 = C_1, \qquad C_0 = H^{-1}, \qquad A_0 = HAH^{-1}, \qquad B_0 = HB \tag{3.39}$$

which apart from a transformation of variable from ξ to $H\xi$ is (3.25).

Equation (3.31) shows that the rate of approach to η to x is determined by the eigenvalues of A_0. In Section 5 it will be shown that these can be assigned arbitrarily under certain conditions. For the analysis given here it was not necessary to assume that the original system (3.22), (3.23) was observable. In Section 5, however, it will be seen that this assumption is necessary if we are to have freedom to assign the eigenvalues of A_0.

The omission of $D(p)u$ from (3.23), where p is the differential operator, does not reduce the generality of our present results. If this term is present, $y - D(p)u$ is used throughout in place of y. The analysis can also be adapted to systems governed by difference equations, but is then less likely to be useful. Such systems will often contain a computer, and then (3.5) can be used directly.

4 Assignment of poles and zeros

Let the equation

$$\dot{x} = Ax + Bu \tag{4.1}$$

be given, with A, B respectively $n \times n$, $n \times l$, and $sI_n - A$, B relatively (left) prime. Let the $m \times n = l \times n$ matrix C in

$$y = Cx \tag{4.2}$$

be subject to choice, and let the resulting system have least order and give transfer function matrix $G(s)$, with McMillan form diag $(\varepsilon_i(s)/\psi_i(s))$. Also, if feedback is applied to the system according to the equation

$$u = v - y \tag{4.3}$$

let the closed-loop transfer function matrix be

$$H(s) = G(s)[I_m + G(s)]^{-1} \tag{4.4}$$

with McMillan form diag $(\varepsilon_i(s)/\psi_i'(s))$. Here we have used the fact (Theorem 1.2) that the numerator polynomials in the McMillan form of H are the same as those in the McMillan form of G.

In this section it is assumed that A, B, C and the coefficients in all polynomials occurring are real, and we shall be concerned with two questions:

(i) Can C be chosen so that the $\varepsilon_i(s)$ in the McMillan form of G are arbitrary monic polynomials satisfying the necessary conditions

(a) ε_i divides ε_{i+1},
(b) $\delta(\varepsilon_1) + \delta(\varepsilon_2) + \cdots + \delta(\varepsilon_m) \leqslant n - m$?

(ii) Can C be chosen so that the $\psi_i'(s)$ in the McMillan form of $H(s)$ are arbitrary monic polynomials satisfying the necessary conditions

(a) ψ_i' divides ψ_{i-1}',
(b) $\delta(\psi_1'\psi_2' \ldots \psi_m') = n$?

In each case the answer is a qualified affirmative, as will be shown in Sections 4.1 and 4.2. The qualification depends partly on the minimal indices of $(sI_n - A\ B)$. It also depends on the need to prevent cancellation in the McMillan forms. As a preliminary step the following lemma is established.

Lemma 4.1 Let a $q \times q$ matrix $Q(s)$ be given, with each element q_{ii} on the principal diagonal monic, of degree d_i, and of higher degree than any other element in the same column. Suppose $d_j < d_k$ for some j, k. Then Q can be transformed by elementary operations to Q' in which each element q_{ii}' on the principal diagonal is monic, has degree d_i', and is of higher degree than any other element in the same column; and moreover

$$\begin{aligned} d_i' &= d_i, \quad i \neq j, k \\ d_j' &= d_j + 1 \\ d_k' &= d_k - 1 \end{aligned} \tag{4.5}$$

Proof Add s times row j to row k. This leaves the degree of each column i (that is, the highest degree of an element in the column) unchanged for $i \neq j, k$. It also leaves unchanged the degrees of the elements in positions (i, j), $i \neq k$, and (i, k), $i \neq k$; places a monic polynomial of degree $d_j + 1$ in position (k, j); and does not increase the degree of the element in position (k, k). Let β be the coefficient (possibly zero) of s^{d_k} in the element in position (k, k). Put $\alpha = d_k - d_j - 1$ and subtract βs^α times column j from column k. This reduces the degree of column k below d_k. Call the resulting matrix Q''. Now construct a $q \times q$ matrix K with constant entries (that is, with entries independent of s) in the following way. Each element k_{rs} of K is equal to the coefficient in q''_{rs} of s^{γ_i}, where $\gamma_i = d_i$, $i \neq j, k$; $\gamma_j = d_j + 1$; $\gamma_k = d_k - 1$. Then if $\delta = d_1 + d_2 + \cdots + d_q$, the coefficient of s^δ in $|Q''(s)|$ is $|K|$, and because $|Q''| = |Q|$, $|K| = 1$ and so K is non-singular. Put $Q'(s) = K^{-1}Q''(s)$. Then it is clear that Q' satisfies the conditions of the theorem.

Example 4.1 Let

$$Q(s) = \begin{pmatrix} s^2+2 & s^3 & s^5+s+1 \\ 2s+1 & s^4+3s+1 & 2s^2 \\ s+2 & 2s^2+1 & s^6-2s^4 \end{pmatrix} \tag{4.6}$$

We shall reduce the degree of the last column, and increase that of the first. The successive stages are:

$$\begin{pmatrix} s^2+2 & s^3 & s^5+s+1 \\ 2s+1 & s^4+3s+1 & 2s^2 \\ s^3+3s+2 & s^4+2s^2+1 & 2s^6-2s^4+s^2+s \end{pmatrix} \tag{4.7}$$

$$Q''(s) = \begin{pmatrix} s^2+2 & s^3 & -s^5-4s^3+s+1 \\ 2s+1 & s^4+3s+1 & -4s^4-2s^3+2s^2 \\ s^3+3s+2 & s^4+2s^2+1 & -8s^4-4s^3+s^2+s \end{pmatrix} \tag{4.8}$$

$$K = \begin{pmatrix} 0 & 0 & -1 \\ 0 & 1 & 0 \\ 1 & 1 & 0 \end{pmatrix}, \qquad K^{-1} = \begin{pmatrix} 0 & -1 & 1 \\ 0 & 1 & 0 \\ -1 & 0 & 0 \end{pmatrix} \tag{4.9}$$

$$Q'(s) = \begin{pmatrix} s^3+s+1 & 2s^2-3s & -4s^4-2s^3-s^2+s \\ 2s+1 & s^4+3s+1 & -4s^4-2s^3+2s^2 \\ -s^2-2 & -s^3 & s^5+4s^3-s-1 \end{pmatrix} \tag{4.10}$$

4.1 Assignment of open-loop zeros

We shall now prove the following theorem.

Theorem 4.1 Let the $n \times n$ matrix A and the $n \times m$ matrix B be given, with $sI_n - A$ and B relatively (left) prime. Let the minimal indices of $(sI_n - A\ B)$ in order of magnitude be $\lambda_1, \lambda_2, \ldots, \lambda_m$, with $\lambda_1 = \lambda_2 = \cdots = \lambda_{m-q} = 0$, $q \leqslant m$. Then the $m \times n$ matrix C may be chosen, relatively (right) prime to $sI_n - A$, so that the McMillan form of $G(s) = C(sI_n - A)^{-1}B$ is diag $(\varepsilon_i(s)/\psi_i(s))$, where

(i) the ψ_i are determined by the Smith form diag $(I_{n-m}, \psi_m, \psi_{m-1}, \ldots, \psi_1)$ of $sI_n - A$, in which $\psi_m = \psi_{m-1} = \cdots = \psi_{t+1} = 1, t \leqslant q$;

(ii) the ε_i may be any prescribed monic polynomials subject to the conditions (of which the first three are necessary)

(a) $\varepsilon_{r+1} = \varepsilon_{r+2} = \cdots = \varepsilon_m = 0$, $t \leqslant r \leqslant q$

(b) ε_i divides ε_{i+1}, $i = 1, 2, \ldots, r - 1$

(c) the degrees $\delta(\varepsilon_i)$ of the non-zero ε_i satisfy

$$\sum_{i=1}^{k} \delta(\varepsilon_i) \leqslant \sum_{i=1}^{k} (\lambda_{m-r+i} - 1), \quad k = 1, 2, \ldots, r \tag{4.11}$$

(d) $r = q$, and ε_q, ψ_1 are relatively prime.

Proof By Exercise 1.5 of Chapter 3, q is the rank of B. If $q < m$, there is a non-singular matrix R, independent of s, such that

$$BR = (B_1 \quad 0_{n,m-q}) \tag{4.12}$$

and the $n \times q$ matrix B_1 has rank q. Also because

$$(sI_n - A \quad B_1, 0) = (sI_n - A \quad B)\begin{pmatrix} I_n & 0 \\ 0 & R \end{pmatrix} \tag{4.13}$$

it follows that $sI_n - A$, B_1 are relatively (left) prime and the minimal indices of $(sI_n - A \quad B_1)$ are $\lambda_{m-q+1}, \lambda_{m-q+2}, \ldots, \lambda_m$.

As in Chapter 3, Section 1.3, we can find constant matrices K_1, K_2 such that $K_1(sI_n - A\ B_1)K_2$ takes the form corresponding to equation (1.49) of Chapter 3,

$$\begin{pmatrix} sI_{p_1} - R_1 & 0 & \cdots & 0 & -E_1 \\ 0 & sI_{p_2} - R_2 & \cdots & 0 & -E_2 \\ \vdots & \vdots & & \vdots & \vdots \\ 0 & 0 & \cdots & sI_{p_q} - R_q & -E_q \end{pmatrix} = (sI_n - A_2 \quad -B_2) \tag{4.14}$$

Here $p_1, p_2, \ldots, p_q$ is some rearrangement of $\lambda_{m-q+1}, \lambda_{m-q+2}, \ldots, \lambda_m$, none of which is zero. The matrices R_i and E_i have the forms given in equations (1.50), (1.51) of Chapter 3. Also if K_2 is partitioned conformably for

$$K_1(sI_n - A \quad B_1)\begin{pmatrix} K_{11} & K_{12} \\ K_{21} & K_{22} \end{pmatrix} = (sI_n - A_2 \quad -B_2) \tag{4.15}$$

it follows on equating coefficients of s that $K_{12} = 0$.

Now adjoin to the matrix $(sI_n - A_2 \quad -B_2)$ a $q \times (n + q)$ matrix $(F\, 0_{q,q})$ to give

$$\begin{pmatrix} sI_n - A_2 & -B_2 \\ F & 0 \end{pmatrix} \tag{4.16}$$

Write F in terms of its columns,

$$F = (f_1, f_2, \ldots, f_n) \tag{4.17}$$

and by column operations eliminate s from rows 1 to n of (4.16). Explicitly, add s times column $n + 1$ to column p_1, then s times column p_1 to column $p_1 - 1, \ldots,$ and then add s times column 2 to column 1. Deal similarly with the other columns. Then by row operations reduce the last q rows to zero except for entries in columns $1, p_1 + 1, \ldots, n - p_q + 1$. Explicitly, do this by adding to rows $n + 1, n + 2, \ldots, n + q$ suitable multiples of rows $1, 2, \ldots, p_1 - 1, p_1 + 1, \ldots, p_1 + p_2 - 1, p_1 + p_2 + 1, \ldots, n - 1$. The final matrix has the form

$$\begin{pmatrix} -R_1 & 0 & \ldots & 0 & -E_1 \\ 0 & -R_2 & \ldots & 0 & -E_2 \\ \vdots & \vdots & & \vdots & \vdots \\ 0 & 0 & \ldots & -R_q & -E_q \\ F_1 & F_2 & \ldots & F_q & 0 \end{pmatrix} \tag{4.18}$$

where

$$F_i = \big(\alpha_i(s), 0, 0, \ldots, 0\big) \tag{4.19}$$

$$\alpha_i(s) = f_{k_i+1} + sf_{k_i+2} + s^2 f_{k_i+3} + \cdots + s^{p_i-1} f_{k_i+p_i} \tag{4.20}$$

$$k_i = p_1 + p_2 + \cdots + p_{i-1} \tag{4.21}$$

Then by changing the signs of the first n rows, and by interchanges of columns, this matrix can be brought to the form,

$$\begin{pmatrix} I_n & 0 \\ 0 & \Phi(s) \end{pmatrix}, \quad \Phi(s) = \big(\alpha_1(s), \alpha_2(s), \ldots, \alpha_q(s)\big) \tag{4.22}$$

The $q \times q$ matrix $\Phi(s)$ has its ith column of degree less than p_i. Let its Smith form be $S(s) = \text{diag}\,(\varepsilon_i(s))$ where $\varepsilon_1, \varepsilon_2, \ldots, \varepsilon_r$ are non-zero, but $\varepsilon_{r+1} = \varepsilon_{r+2} =$

$\cdots = \varepsilon_q = 0$. Then because $\varepsilon_1\varepsilon_2\ldots\varepsilon_i$ is the greatest common divisor of minors of order i in $\Phi(s)$, it follows that

$$\left.\begin{aligned} \delta(\varepsilon_1) &\leqslant \lambda_{m-r+1} - 1 \\ \delta(\varepsilon_1\varepsilon_2) &\leqslant \lambda_{m-r+1} + \lambda_{m-r+2} - 2 \\ &\vdots \\ \delta(\varepsilon_1\varepsilon_2\ldots\varepsilon_r) &\leqslant \lambda_{m-r+1} + \lambda_{m-r+2} + \cdots + \lambda_m - r \end{aligned}\right\} \tag{4.23}$$

Subject to these conditions, which are clearly necessary, the monic polynomials $\varepsilon_i(s)$ can be assigned arbitrarily by choice of F. This is proved in the following way.

If the given ε_i are such that

$$\delta(\varepsilon_i) \leqslant \lambda_{m-r+i} - 1, \quad i = 1, 2, \ldots, r, \quad \varepsilon_{r+1} = \varepsilon_{r+2} = \cdots = \varepsilon_q = 0 \tag{4.24}$$

then S needs only a change of order of the ε_i to qualify as Φ. If not, suppose

$$\delta(\varepsilon_k) > \lambda_{m-r+k} - 1, \quad k \leqslant r \tag{4.25}$$

Then because

$$\sum_{i=1}^{k} [\delta(\varepsilon_i) - (\lambda_{m-r+i} - 1)] \leqslant 0 \tag{4.26}$$

there is some $j < k$ for which

$$\delta(\varepsilon_j) < \lambda_{m-r+j} - 1 \leqslant \lambda_{m-r+k} - 1 < \delta(\varepsilon_k) \tag{4.27}$$

As in Lemma 4.1, the degree of the column j in S can be increased by 1, and the degree of column k reduced by 1. By successive operations of this kind, the degree of column i can be made less than λ_{m-r+i}, $i = 1, 2, \ldots, r$. Then by an interchange of columns, the degree of column i is made less than p_i.

Given ε_i as in the theorem, we may therefore choose F so that the Smith form of Φ is diag $(\varepsilon_1, \varepsilon_2, \ldots, \varepsilon_q)$. Then because (4.22) was obtained from (4.16) by elementary operations, the Smith form of

$$\begin{pmatrix} sI_n - A_2 & -B_2, 0_{n,m-q} \\ \hline F & \\ & 0_{m,m} \\ 0_{m-q,n} & \end{pmatrix} \tag{4.28}$$

is

$$\begin{pmatrix} I_n & 0 & 0 \\ 0 & \text{diag}\,(\varepsilon_1, \varepsilon_2, \ldots, \varepsilon_q) & 0 \\ 0 & 0 & 0_{m-q,m-q} \end{pmatrix} \tag{4.29}$$

By (4.13) and (4.15), this is also the Smith form of

$$P_0(s) = \begin{pmatrix} sI_n - A & B \\ -C & 0 \end{pmatrix} \tag{4.30}$$

$$= \begin{pmatrix} K_1^{-1} & 0 \\ 0 & I_m \end{pmatrix} \left(\begin{array}{c|c} sI_n - A_2 & -B_2, 0 \\ \hline F & 0 \\ 0 & \end{array}\right)$$

$$\times \left(\begin{array}{c|cc} K_{11}^{-1} & 0 & 0 \\ \hline -K_{22}^{-1}K_{21}K_{11}^{-1} & K_{22}^{-1} & 0 \\ 0 & 0 & I_{m-q} \end{array}\right) \begin{pmatrix} I_n & 0 \\ 0 & R^{-1} \end{pmatrix} \tag{4.31}$$

in which C is given explicitly by

$$C = \begin{pmatrix} FK_{11}^{-1} \\ 0 \end{pmatrix} \tag{4.32}$$

and $K_{11}^{-1} = K_1 = H_1$, where H_1 is a matrix of the type defined in equation (1.22) of Chapter 3, but derived from $sI_n - A$ and B_1 [compare Chapter 3, equation (1.48)].

If $P_0(s)$ has an o.d. zero and $r = q$, then because

$$\begin{pmatrix} sI_n - A & B \\ -C & 0 \end{pmatrix} = \left(\begin{array}{c|cc} sI_n - A & B_1 & 0 \\ \hline -FK_{11}^{-1} & 0 & 0 \\ 0 & 0 & 0 \end{array}\right) \begin{pmatrix} I_n & 0 \\ 0 & R^{-1} \end{pmatrix} \tag{4.33}$$

it follows that there is some η for which

$$|\eta I_n - A| = 0 \tag{4.34}$$

$$\begin{vmatrix} \eta I_n - A & B_1 \\ -FK_{11}^{-1} & 0 \end{vmatrix} = 0 \tag{4.35}$$

But this means that η is a zero of ψ_1 and also of ε_q, which is impossible by condition (iid) of the theorem, and so it follows that P_0 has least order.

By Theorem 4.1 of Chapter 3, the McMillan form of the transfer function matrix derived from P has numerator polynomials $\varepsilon_1(s), \varepsilon_2(s), \ldots, \varepsilon_q(s)$, followed by $m - q$ zeros. The denominator polynomials are the last m invariant polynomials of $sI_n - A$, with their order reversed. Because $sI_n - A$, B_1 are relatively (left) prime by (4.33), there can be no more than q non-unit polynomials among these m invariant polynomials. This concludes the proof.

Corollary Let A, B, λ_i, q be as in the statement of the theorem. Then the $m \times n$ matrix C may be chosen so that the Smith form of the system matrix $P_0(s)$ given in equation (4.30) is diag $(I_n, \varepsilon_1, \varepsilon_2, \ldots, \varepsilon_m)$. Here the ε_i may be any prescribed monic polynomials satisfying the necessary and sufficient conditions (iia), (iib), (iic) of the theorem.

Proof This was established in the proof at equation (4.30) without use of condition (iid).

EXERCISE 4.1 Devise an alternative proof using Lemma 4.1, and (ii) of Exercise 2.4, Chapter 3.

4.2 Assignment of closed-loop poles

The second question posed at the beginning of Section 4 is answered by the following theorem.

Theorem 4.2 With A, B, and $\lambda_1, \lambda_2, \ldots, \lambda_m$ as in Theorem 4.1, the $m \times n$ matrix C may be chosen, relatively (right) prime to $sI_n - A$, so that the McMillan form of $H(s) = G(s)[I_m + G(s)]^{-1}$ is diag $(\varepsilon_i(s)/\psi_i'(s))$, where

(i) the ε_i are the numerator polynomials in the McMillan form diag $(\varepsilon_i(s)/\psi_i(s))$ of $G(s)$;

(ii) the ψ' are any prescribed non-zero monic polynomials subject to the conditions (of which the first two are necessary);

(a) for $i = 2, 3, \ldots, m$, ψ_i' divides ψ_{i-1}',

(b) the degrees $\delta(\psi_i')$ of the ψ_i' satisfy

$$\sum_{i=1}^{k} \delta(\psi'_{m+1-i}) \leqslant \sum_{i=1}^{k} \lambda_i, \quad k = 1, 2, \ldots, m \tag{4.36}$$

with equality holding for $k = m$,

(c) ψ_1' and ψ_1 are relatively prime.

Proof When feedback is applied according to equation (4.3), the system equations become

$$(sI_n - A + BC)\bar{x} = B\bar{v} \tag{4.37}$$

$$\bar{y} = C\bar{x} \tag{4.38}$$

Because $sI_n - A$, B are relatively (left) prime and

$$(sI_n - A + BC \quad B) = (sI_n - A \quad B)\begin{pmatrix} I_n & 0 \\ C & I_m \end{pmatrix} \tag{4.39}$$

it follows (compare Exercise 1.3) that $sI_n - A + BC$ and B are relatively (left) prime. In the same way, if C (which is still to be chosen) is relatively (right) prime to $sI_n - A$, it will be relatively (right) prime to $sI_n - A + BC$. Then by Theorem 1.2, the McMillan forms of $H(s)$ and $G(s)$ will share the same numerator polynomials ε_i. The denominator polynomials ψ_i' in the McMillan form of $H(s)$, by Theorem 1.2, will be the last m diagonal entries (in reversed order) in the Smith form of

$$\begin{pmatrix} sI_n - A & B & 0 \\ -C & 0 & -I_m \\ 0 & I_m & I_m \end{pmatrix} \tag{4.40}$$

which is the same as the Smith form of

$$\begin{pmatrix} I_m & 0 & 0 \\ 0 & sI_n - A & B \\ 0 & -C & I_m \end{pmatrix} \tag{4.41}$$

Now proceed as in the proof of Theorem 4.1, but in (4.16) adjoin the matrix $(F\ I_q)$ to give

$$\begin{pmatrix} sI_n - A_2 & -B_2 \\ F & I_q \end{pmatrix} \tag{4.42}$$

Write I_q in terms of its columns,

$$I_q = (e_1, e_2, \ldots, e_q) \tag{4.43}$$

Then with exactly the same operations as before a matrix

$$\begin{pmatrix} -R_1 & 0 & \cdots & 0 & -E_1 \\ 0 & -R_2 & \cdots & 0 & -E_2 \\ \vdots & \vdots & & \vdots & \vdots \\ 0 & 0 & \cdots & -R_q & -E_q \\ F_1' & F_2' & \cdots & F_q' & I_q \end{pmatrix} \tag{4.44}$$

is obtained instead of (4.18). Here

$$F_i' = \big(\alpha_i'(s), 0, 0, \ldots, 0\big) \tag{4.45}$$

$$\alpha_i'(s) = f_{k_i+1} + sf_{k_i+2} + \cdots + s^{p_i-1}f_{k_i+p_i} + s^{p_i}e_i \tag{4.46}$$

with k_i as in (4.21). The submatrix I_q at the bottom right of (4.44) can be reduced to zero by adding row p_1 to row $n + 1$, row $(p_1 + p_2)$ to row $n + 2$, and so on. The invariant polynomials $\psi_1', \psi_2', \ldots, \psi_q'$ of

$$\Phi'(s) = \big(\alpha_1'(s), \alpha_2'(s), \ldots, \alpha_q'(s)\big) \tag{4.47}$$

must clearly obey condition (iib) of the theorem. As in the proof of Theorem 4.1, the matrix F may be chosen so that the monic polynomials have any prescribed form subject to the conditions of the theorem. The remaining invariant polynomials of $H(s)$, $\psi'_{q+1}, \psi'_{q+2}, \ldots, \psi'_m$, must be 1 because $\lambda_1, \lambda_2, \ldots, \lambda_{m-q}$ have been assumed zero.

Having chosen F appropriately, let $Q = R\begin{pmatrix} K_{22} & 0 \\ 0 & I_{m-q} \end{pmatrix}$ and obtain from (4.42)

$$P_1(s) = \begin{pmatrix} sI_n - A & B \\ -C & I_m \end{pmatrix}$$

$$= \begin{pmatrix} K_1^{-1} & 0 \\ 0 & Q \end{pmatrix} \left(\begin{array}{c|cc} sI_n - A_2 & -B_2 & 0 \\ \hline F & I_q & 0 \\ 0 & 0 & I_{m-q} \end{array}\right)$$

$$\times \left(\begin{array}{c|cc} K_{11}^{-1} & 0 & 0 \\ \hline -K_{22}^{-1}K_{21}K_{11}^{-1} & K_{22}^{-1} & 0 \\ 0 & 0 & I_{m-q} \end{array}\right) \begin{pmatrix} I_n & 0 \\ 0 & R^{-1} \end{pmatrix} \tag{4.48}$$

in which the Smith form of P_1 is diag $(I_n, \psi'_m, \psi'_{m-1}, \ldots, \psi'_1)$. If P_1 had an o.d. zero, ψ'_1 and ψ_1 would have a common zero, which is excluded by condition (iic). Therefore P_1 has least order, and the theorem follows.

Corollary 1 Let A, B, λ_i be as in the statement of the theorem. Then C can be chosen so that the Smith form of the matrix $sI_n - A + BC$ is

$$\text{diag}\,(I_{n-m}, \psi'_m, \psi'_{m-1}, \ldots, \psi'_1)$$

Here the ψ' may be any monic polynomials satisfying the necessary and sufficient conditions (iia), (iib).

Proof The matrix P_1 in (4.48) can be brought by elementary operations to the form

$$\begin{pmatrix} I_m & 0 \\ 0 & sI_n - A + BC \end{pmatrix} \tag{4.49}$$

Assumption (iic) is not required.

Corollary 2 If $sI_n - A$, B are relatively (left) prime, and $\phi(s)$ is any monic polynomial of degree n, there exists an $m \times n$ matrix C such that

$$|sI_n - A + BC| = \phi(s) \tag{4.50}$$

Proof Choose $\psi'_1 = \phi$, $\psi'_2 = \psi'_3 = \cdots = \psi'_m = 1$. Conditions (iia), (iib) are fulfilled, and the result follows from Corollary 1.

Condition (iic) is not needed in the corollaries, but without some such condition $sI_n - A$ and C need not be relatively (right) prime. Then cancellation will occur in G and the corresponding poles will be absent from H.

Example 4.2 Let $\phi(s) = (s + 1)(s + 2)^2$ and

$$(sI_3 - A \quad B) = \begin{pmatrix} s & 0 & 1 & 1 \\ -1 & s & 3 & 0 \\ 0 & -1 & s+3 & 0 \end{pmatrix} \tag{4.51}$$

so that $\psi_1(s) = (s + 1)^3$. Then elementary calculations give $C = (2, -1, 0)$, so that

$$|sI_3 - A + BC| = \begin{vmatrix} s+2 & -1 & 1 \\ -1 & s & 3 \\ 0 & -1 & s+3 \end{vmatrix} = (s + 1)(s + 2)^2 \tag{4.52}$$

but

$$G(s) = (2 \quad -1 \quad 0)\begin{pmatrix} s & 0 & 1 \\ -1 & s & 3 \\ 0 & -1 & s+3 \end{pmatrix}^{-1}\begin{pmatrix} 1 \\ 0 \\ 0 \end{pmatrix}$$

$$= \frac{(s + 1)(2s + 3)}{(s + 1)^3} \tag{4.53}$$

$$H(s) = \frac{G}{1 + G} = \frac{2s + 3}{(s + 2)^2} \tag{4.54}$$

Here $sI_3 - A$, C are not relatively (right) prime, and the pole at -1 is absent from $H(s)$.

Example 4.3 Condition (iic) is not universally necessary, as can be seen by the example

$$A = \begin{pmatrix} -1 & 0 \\ 0 & -2 \end{pmatrix}, \quad B = \begin{pmatrix} 1 & 0 \\ 0 & 1 \end{pmatrix},$$

$$|sI_2 - A + BC| = (s + 1)(s + 2) \tag{4.55}$$

which can be achieved with $C = 0$, but also with

$$C = \begin{pmatrix} 1 & 0 \\ 0 & -1 \end{pmatrix} \tag{4.56}$$

giving

$$G(s) = \begin{pmatrix} 1/(s+1) & 0 \\ 0 & -1/(s+2) \end{pmatrix} \tag{4.57}$$

$$H(s) = \begin{pmatrix} 1/(s+2) & 0 \\ 0 & -1/(s+1) \end{pmatrix} \tag{4.58}$$

Here $\psi_1(s) = (s+1)(s+2) = \psi_1'(s)$, yet $sI_2 - A$, C are relatively (right) prime.

All the results which have been given in Sections 4.1 and 4.2 for $sI_n - A$, B translate at once into theorems concerning $sI_n - A$, C. The easiest way to obtain these is to consider $sI_n - A^{\mathrm{T}}$, C^{T}.

4.3 Relatively prime polynomial matrices

The results in Sections 4.1 and 4.2 allow us to give, for the special case where $T(s) = sI_n - A$, $U(s) = B$, a sharper form of the result in equation (6.11) of Chapter 2.

Theorem 4.3 Let $sI_n - A$, B be relatively (left) prime, and let B have rank q. Then there exists a unimodular $(n + l) \times (n + l)$ matrix $N(s)$ such that

$$(sI_n - A \quad B)N(s) = (sI_n - A \quad B)\begin{pmatrix} N_{11}(s) & N_{12}(s) \\ N_{21}(s) & N_{22}(s) \end{pmatrix} \tag{4.59}$$

$$= (I_n \quad 0) \tag{4.60}$$

where in $N(s)$ the highest degree of an element

(i) in $N_{11}(s)$ is $n - q - 1$,
(ii) in $N_{12}(s)$ or $N_{21}(s)$ is $n - q$,
(iii) in $N_{22}(s)$ is $n - q + 1$.

Proof Define an $n \times q$ matrix B_1 with rank q by

$$BR = (B_1 \quad 0) \tag{4.61}$$

where R is a constant matrix with $|R| = 1$. It is easy to see that $sI_n - A$, B_1 are relatively (left) prime. Choose the $q \times n$ matrix C_1 so that

$$\begin{vmatrix} sI_n - A & B_1 \\ -C_1 & 0 \end{vmatrix} = 1 \tag{4.62}$$

Then

$$\begin{vmatrix} sI_n - A & B \\ -C & D \end{vmatrix} = \left| \begin{pmatrix} sI_n - A & B_1 & 0 \\ -C_1 & 0 & 0 \\ 0 & 0 & I_{l-q} \end{pmatrix} \begin{pmatrix} I_n & 0 \\ 0 & R^{-1} \end{pmatrix} \right| = 1 \tag{4.63}$$

Also

$$(I_n \quad 0)\begin{pmatrix} sI_n - A & B \\ -C & D \end{pmatrix} = (sI_n - A \quad B) \tag{4.64}$$

whence the second matrix on the left-hand side qualifies as $N^{-1}(s)$. The theorem now follows from Laplace's expansion of the $(n + l - 1) \times (n + l - 1)$ minors of N^{-1}.

Corollary 1 A necessary and sufficient condition for $sI_n - A$, B to be relatively (left) prime is that there exist polynomial matrices $X(s)$, $Y(s)$, having their elements of degree no higher than $n - q - 1$, $n - q$ respectively, such that

$$(sI_n - A)X(s) + BY(s) = I_n \tag{4.65}$$

[Compare Chapter 2, equation (6.20).]

Proof With N defined as in (4.59),

$$(sI_n - A \quad B)\begin{pmatrix} N_{11}(s) \\ N_{21}(s) \end{pmatrix} = I_n \tag{4.66}$$

and $X = N_{11}$, $Y = N_{21}$ satisfy (4.65).

Corollary 2 Let $sI_n - A$, B be relatively (left) prime, and $sI_n - A_1$, B_1 be also relatively (left) prime, where B and B_1 are both $n \times l$. Then there exists an $(n + l) \times (n + l)$ unimodular matrix $R(s)$, having every element of degree not exceeding n, such that

$$(sI_n - A_1 \quad B_1) = (sI_n - A \quad B)R(s) \tag{4.67}$$

Proof As in (4.64),

$$(sI_n - A_1 \quad B_1) = (I_n \quad 0)\begin{pmatrix} sI_n - A_1 & B_1 \\ -C_1 & D_1 \end{pmatrix} \tag{4.68}$$

$$= (sI_n - A \quad B)\begin{pmatrix} sI_n - A & B \\ -C & D \end{pmatrix}^{-1}\begin{pmatrix} sI_n - A_1 & B_1 \\ -C_1 & D_1 \end{pmatrix} \tag{4.69}$$

$$= (sI_n - A \quad B)R(s) \tag{4.70}$$

whence the result follows easily.

5 Observers with arbitrary dynamics

In Section 3.1, no results were given concerning the location of the eigenvalues of A_0. This can now be done: instead of deriving H, C_0, C_1, A_0, B_1, B_0 by the method given in Section 3.1, proceed in the following way.

Assume as before that $sI_n - A$, C are relatively (right) prime and that C has rank m. There then exists a non-singular $n \times n$ matrix H_1 such that

$$CH_1 = (0 \quad I_m) \tag{5.1}$$

whence

$$\begin{pmatrix} H_1^{-1} & 0 \\ 0 & I_m \end{pmatrix}\begin{pmatrix} sI_n - A & B \\ -C & 0 \end{pmatrix}\begin{pmatrix} H_1 & 0 \\ 0 & I_m \end{pmatrix}$$

$$= \left(\begin{array}{cc|c} sI_{n-m} - A_{11} & -A_{12} & B_1'' \\ -A_{21} & sI_m - A_{22} & B_2'' \\ \hline 0 & -I_m & 0 \end{array}\right) \tag{5.2}$$

Because the last matrix has no o.d. zeros, its first $n - m$ columns have rank $n - m$ for all s. Hence $sI_{n-m} - A_{11}$, A_{21} are relatively (right) prime. Put

$$H_2 = \begin{pmatrix} I_{n-m} & 0 \\ A_{21} & I_m \end{pmatrix} \tag{5.3}$$

and carry out a further transformation of system similarity on the last matrix in (5.2) with transforming matrix H_2. The result is

$$\left(\begin{array}{cc|c} sI_{n-m} - A_{11} - A_{12}A_{21} & -A_{12}' & B_1' \\ -A_{21}' & sI_m - A_{22}' & B_2' \\ \hline -A_{21} & -I_m & 0 \end{array}\right) \tag{5.4}$$

which we write

$$\begin{pmatrix} sI_n - A' & B' \\ -C' & 0 \end{pmatrix} \tag{5.5}$$

Now regard (5.5) as the given system, and choose

$$H = (I_{n-m} \quad 0) \tag{5.6}$$

which ensures that the matrix

$$\begin{pmatrix} H \\ C' \end{pmatrix} \tag{5.7}$$

is certainly non-singular. Because $(HH^{\mathrm{T}}) = I_{n-m}$, equation (3.30) gives

$$A_0 = HA'H^{\mathrm{T}} - B_1C'H^{\mathrm{T}} \tag{5.8}$$

where from (5.4)

$$sI_{n-m} - HA'H^{\mathrm{T}} = sI_{n-m} - A_{11} - A_{12}A_{21} \tag{5.9}$$

$$C'H^{\mathrm{T}} = A_{21} \tag{5.10}$$

and it readily follows (compare Exercise 1.3) that these are relatively (right) prime. By Corollary 2 of Theorem 4.2, suitably modified, an $(n - m) \times m$ matrix B_1 can be chosen so that

$$|sI_{n-m} - A_0| = \phi(s) \tag{5.11}$$

where $\phi(s)$ is any monic polynomial of degree $n - m$.

Now obtain B_0, C_0, C_1 from

$$B_0 = HB' \tag{5.12}$$

$$(C_0 \quad C_1) = H_1H_2\begin{pmatrix} H \\ C' \end{pmatrix}^{-1} \tag{5.13}$$

It follows from (3.28) and (3.32) that if the roots of $\phi(s)$ have negative real parts, η tends as $t \to \infty$ to H_1H_2x', where x' is the state vector for the system (5.5). Since $H_1H_2x' = x$, the state vector for the original system, it follows that η provides the required estimate of x.

Finally, suppose that a system is given, as above,

$$\dot{x} = Ax + Bu \tag{5.14}$$

$$y = Cx \tag{5.15}$$

and that an observer of order $n - m$ has been derived

$$\dot{\xi} = A_0\xi + B_0u + B_1y \tag{5.16}$$

$$\eta = C_0\xi + C_1y \tag{5.17}$$

Let η be used instead of x to provide negative feedback according to the equation

$$u = -C_2\eta \tag{5.18}$$

Then putting $e = \xi - Hx$, equation (3.31) becomes

$$\dot{e} = A_0e \tag{5.19}$$

while (3.33) leads to

$$\eta = C_0e + x \tag{5.20}$$

so that (5.14) becomes

$$\dot{x} = (A - BC_2)x - BC_2C_0e \tag{5.21}$$

Equations (5.19) and (5.21) show that the eigenvalues of the closed-loop system are those of A_0 and those of $A - BC_2$. It has already been shown that $|sI_{n-m} - A_0|$ can be made equal to any given monic polynomial $\phi(s)$ of degree $n - m$, provided that $sI_n - A$, C are relatively (right) prime. If $sI_n - A$, B are relatively (left) prime, C_2 can also be chosen so that $|sI_n - A + BC_2|$ is equal to any given monic polynomial $\psi(s)$ of degree n. Therefore if the system (5.14), (5.15) has least order, all $2n - m$ eigenvalues of the closed-loop system can be assigned. When this is done, however, it should be noticed that $\bar{u}$ is related to $\bar{y}$ by

$$\bar{u} = \{-C_2C_0[sI_{n-m} - A_0 + B_0C_2C_0]^{-1}(B_1 - B_0C_2C_1) - C_2C_1\}\bar{y} \tag{5.22}$$

The stability, or the amount of phase advance, provided by the controller defined in (5.22) may fail to be satisfactory.

6 Closed-loop stability

This section will be concerned with the stability of systems to which feedback is applied. The emphasis will be on tests which are easy to apply to practical systems.

6.1 Systems described by differential equations

Let feedback be applied to the system

$$T(s)\bar{\xi} = U(s)\bar{u} \tag{6.1}$$

$$\bar{y} = V(s)\bar{\xi} + W(s)\bar{u} \tag{6.2}$$

in which $l = m$, according to the equation

$$\bar{u} = \bar{v} - F(s)\bar{y} \tag{6.3}$$

as in Section 1, where $F(s)$ is a polynomial matrix. Write

$$\phi_0(s) = |T(s)| \tag{6.4}$$

$$\phi_1(s) = \begin{vmatrix} T(s) & U(s) \\ -V(s) & W(s) \end{vmatrix} \tag{6.5}$$

$$\phi_2(s) = \begin{vmatrix} T(s) & U(s) & 0 \\ -V(s) & W(s) & -I_m \\ 0 & I_m & F(s) \end{vmatrix} \tag{6.6}$$

Then if $G(s)$, $H(s)$ are respectively the open- and closed-loop transfer function matrices, it follows as in Section 1 that

$$|G(s)| = \frac{\phi_1(s)}{\phi_0(s)} \tag{6.7}$$

$$|H(s)| = \frac{\phi_1(s)}{\phi_2(s)} \tag{6.8}$$

and so, as in (1.24),

$$\frac{|G(s)|}{|H(s)|} = \frac{\phi_2(s)}{\phi_0(s)} \tag{6.9}$$

The closed-loop system is asymptotically stable if and only if all the zeros of $\phi_2(s)$ lie in the open left half-plane, as follows at once from the system matrix $P_c(s)$ in (1.6). For the purposes of design it is easier to work with G and H, or $\hat{G} = G^{-1}$ and $\hat{H} = H^{-1} = \hat{G} + F$. To deduce stability from these it is necessary to recognize that cancellation may occur in (6.9). Extra conditions are therefore needed to ensure that the cancelled poles are not unstable; the simplest condition of this type is to assume that the open-loop system (6.1) is asymptotically stable. This is the course which will be adopted here.

In what follows we shall need a contour D in the complex plane which is defined as follows. Let D consist of the imaginary axis from $-iR$ to $+iR$, together with a semicircle of radius R about the origin in the right half-plane. Let R be chosen so that the semicircle does not intercept any pole or zero of $|G|$ or $|H|$. If any poles or zeros of $|G|$ or $|H|$ lie on the imaginary axis between $-iR$ and $+iR$, let D be indented into the left half-plane to avoid them. The indentations are to be small enough to avoid any pole or zero in the open left half-plane, which is possible because $|G|$ and $|H|$ are rational functions with isolated poles and zeros. Finally, let R be chosen large enough to ensure that every finite pole and zero of $|G|$ and $|H|$ lying in the closed right half-plane is inside D.

Theorem 6.1 Let (6.1) be asymptotically stable. Let $|G|$ map D into Γ_0, and $|H|$ map D into Γ_c. Then the closed-loop system is asymptotically stable if and only if Γ_c, Γ_0 encircle the origin the same number of times. (Traditionally D is described clockwise, and clockwise encirclements of the origin are counted positive.)

Proof Let $\phi_0(s)$ have z_0 zeros in the closed right half-plane; because (6.1) is asymptotically stable $z_0 = 0$. Similarly, let $\phi_1(s)$, $\phi_2(s)$ have z_1, z_2 zeros in the closed right half-plane. Let Γ_0, Γ_c encircle the origin N_0, N_c times. Then regardless of any cancellation in (6.7), (6.8), the 'principle of the argument' gives

$$N_0 = z_1 - z_0 = z_1 \tag{6.10}$$

$$N_c = z_1 - z_2 \tag{6.11}$$

The closed-loop system is asymptotically stable if and only if $z_2 = 0$, that is, by (6.10), (6.11), if and only if $N_c = N_0$.

Corollary Let (6.1) be asymptotically stable. Let $|\hat{G}| = |G|^{-1}$ map D into $\hat{\Gamma}_0$, and $|\hat{H}| = |H|^{-1}$ map D into $\hat{\Gamma}_c$. Then the closed-loop system is asymptotically stable if and only if $\hat{\Gamma}_c$, $\hat{\Gamma}_0$ encircle the origin the same number of times.

Example 6.1 Let $F = I_2$, and

$$G = \begin{pmatrix} \dfrac{1}{s+1} & 0 \\ 0 & \dfrac{1}{s+2} \end{pmatrix} \tag{6.12}$$

giving

$$H = \begin{pmatrix} \dfrac{1}{s+2} & 0 \\ 0 & \dfrac{1}{s+3} \end{pmatrix} \tag{6.13}$$

$$|G| = \frac{1}{(s+1)(s+2)} \tag{6.14}$$

$$|H| = \frac{1}{(s+2)(s+3)} \tag{6.15}$$

$$\frac{|G|}{|H|} = \frac{s+3}{s+1} \tag{6.16}$$

Neither Γ_c nor Γ_0 encircles the origin, and provided that G arises from an asymptotically stable system, the closed-loop system is also asymptotically stable. Notice that if the system giving rise to G has a decoupling zero, then the system giving rise to H will have the same decoupling zero, by Theorem 1.1. The condition that the open-loop system shall be asymptotically stable is necessary to ensure that such decoupling zeros do not destroy the asymptotic stability of H. Notice also that the pole at -2 of G cancels with the pole at -2 of H in $|G|/|H|$.

The preceding theorem is difficult to apply in general, because $|H|$, or even $|\hat{H}| = |F + \hat{G}|$ is a complicated function of F. A more convenient result can be obtained by means of the following theorem. First let $Q(s)$ be a rational $m \times m$ matrix and define

$$d_i(s) = \sum_{\substack{j=1 \\ j \neq i}}^{m} |q_{ij}(s)| \tag{6.17}$$

$$\delta_i(s) = \sum_{\substack{j=1 \\ j \neq i}}^{m} |q_{ji}(s)| \tag{6.18}$$

Define D in a similar way to that used before so that no pole or zero of $|Q(s)|$ or $q_{ij}(s)$, $i, j = 1, 2, \ldots, m$ lies on D, and every finite pole and zero of $|Q(s)|$ and $q_{ii}(s)$, $i = 1, 2, \ldots, m$, lying in the closed right half-plane is inside D. Let $|Q(s)|$ map D into Γ and $q_{ii}(s)$ map D into Γ_i. Let Γ encircle the origin N times, and Γ_i encircle the origin N_i times.

Theorem 6.2 For all s on D, let $|q_{ii}(s)| - d_i(s) > 0$ (resp. $|q_{ii}(s)| - \delta_i(s) > 0$), $i = 1, 2, \ldots, m$. Then $N = N_1 + N_2 + \cdots + N_m$.

Proof Let $Q(\theta, s)$ be the matrix having

$$q_{ii}(\theta, s) = q_{ii}(s) \tag{6.19}$$

$$q_{ij}(\theta, s) = \theta q_{ij}(s), \quad j \neq i \tag{6.20}$$

where $q_{ii}(s)$, $q_{ij}(s)$ are the elements of $Q(s)$ and $0 \leqslant \theta \leqslant 1$. By the construction of D, $Q(\theta, s)$ is a finite matrix for all s on D and all $\theta \in [0, 1]$. Consider the function

$$\phi(\theta, s) = \frac{|Q(\theta, s)|}{\prod_{i=1}^{m} q_{ii}(s)} \tag{6.21}$$

which is finite for all $0 \leqslant \theta \leqslant 1$ and all s on D, and gives

$$\phi(0, s) = 1 \tag{6.22}$$

Let $\phi(1, s)$ map D into Γ_0. For given s on D, $\phi(\theta, s)$ defines a continuous curve $\gamma(s)$ joining the point $(1, 0)$ in the complex plane to the point on Γ_0 corresponding to s. As s traces out D, starting from $s = 0$, so $\gamma(s)$ sweeps out a region of the complex plane and returns at last to its original position.

Suppose, contrary to what is to be proved, that Γ_0 encircles the origin. Then the region swept out by $\gamma(s)$ as s traverses D includes the origin. That is, there is some $\theta \in [0, 1]$, and some s on D, for which $\phi(\theta, s) = 0$. But the $q_{ii}(s)$ are all finite on D, and so $Q(\theta, s) = 0$. By Gershgorin's theorem (Chapter 1, Theorem 1.9) this is impossible. Then from (6.21) the number of encirclements of the origin by Γ_0 is

$$0 = N - N_1 - N_2 - \cdots - N_m \tag{6.23}$$

which proves the theorem.

This theorem can be applied in several ways. Generally it is easier to work with $\hat{G}$ and $\hat{H} = F + \hat{G}$ than with G and H. Let D be as before, with no pole or zero of $|\hat{G}|$, $|\hat{H}|$, $\hat{g}_{ij}$, $\hat{h}_{ij}$ on D, and with every finite pole and zero of $|\hat{G}|$, $|\hat{H}|$, $\hat{g}_{ii}$, $\hat{h}_{ii}$ which is in the closed right half-plane lying inside D. Let D be mapped by $|\hat{G}|$ into $\hat{\Gamma}_0$, by $|\hat{H}|$ into $\hat{\Gamma}_c$, by $\hat{g}_{ii}$ into $\hat{\Gamma}_{oi}$, and by $\hat{h}_{ii}$ into $\hat{\Gamma}_{ci}$. Let the number of encirclements of the origin by $\hat{\Gamma}_0$ be N_0, by $\hat{\Gamma}_c$ be N_c, by $\hat{\Gamma}_{oi}$ be N_{oi}, and by $\hat{\Gamma}_{ci}$ be N_{ci}. Define $d_i(s)$ and $\delta_i(s)$ by

$$d_i(s) = \sum_{\substack{j=1 \\ j \neq i}}^{m} |\hat{g}_{ij}(s)| \tag{6.24}$$

$$\delta_i(s) = \sum_{\substack{j=1 \\ j \neq i}}^{m} |\hat{g}_{ji}(s)| \tag{6.25}$$

Theorem 6.3 Let the open-loop system (6.1) be asymptotically stable and let $|G|$ have no zero in the closed right half-plane. Let F be diagonal and for all s on D let $|f_{ii}(s) + \hat{g}_{ii}(s)| - d_i(s) > 0$ (resp. $|f_{ii}(s) + \hat{g}_{ii}(s)| - \delta_i(s) > 0$), $i = 1, 2, \ldots, m$. Then the closed-loop system is asymptotically stable if and only if

$$\sum_{i=1}^{m} N_{ci} = 0 \tag{6.26}$$

Proof Because (6.1) is asymptotically stable, $|\hat{G}|$ has no zero in the closed right half-plane: it also has no pole there. Consequently $N_0 = 0$. By the corollary of Theorem 6.1, the closed-loop system is asymptotically stable if and only if $N_c = 0$. By Theorem 6.2,

$$N_c = \sum_{i=1}^{m} N_{ci} \tag{6.27}$$

which completes the proof.

Corollary A sufficient condition for the asymptotic stability of the closed-loop system, subject to the other conditions of the theorem, is obtained by replacing (6.26) by

$$N_{ci} = 0, \quad i = 1, 2, \ldots, m \tag{6.28}$$

This is the usual inverse Nyquist criterion applied separately to each of the $\hat{g}_{ii}$.

In Theorem 6.3, the condition that F is diagonal is very easily relaxed. All that is needed is to replace $\hat{g}_{ij}(s)$ in (6.24), (6.25) by $f_{ij}(s) + \hat{g}_{ij}(s)$. A stronger result than Theorem 6.3 can be obtained if Theorem 6.2 is applied twice.

Theorem 6.4 Let the open-loop system (6.1) be asymptotically stable. Let F be diagonal and for all s on D and $i = 1, 2, \ldots, m$ let

(i) $|\hat{g}_{ii}(s)| - d_i(s) > 0$ (resp. $|\hat{g}_{ii}(s)| - \delta_i(s) > 0$);
(ii) $|f_{ii}(s) + \hat{g}_{ii}(s)| - d_i(s) > 0$ (resp. $|f_{ii}(s) + \hat{g}_{ii}(s)| - \delta_i(s) > 0$).

Then the closed-loop system is asymptotically stable if and only if

$$\sum_{i=1}^{m} N_{ci} = \sum_{i=1}^{m} N_{oi} \tag{6.29}$$

Proof By Theorem 6.1, the closed-loop system is asymptotically stable if and only if $N_c = N_o$. By Theorem 6.2, applied twice

$$N_0 = \sum_{i=1}^{m} N_{oi}, \qquad N_c = \sum_{i=1}^{m} N_{ci} \tag{6.30}$$

which completes the proof.

Corollary A sufficient condition for the asymptotic stability of the closed-loop system, subject to the other conditions of the theorem, is obtained by replacing (6.29) by

$$N_{ci} = N_{oi}, \quad i = 1, 2, \ldots, m \tag{6.31}$$

This is the usual inverse Nyquist criterion applied separately to each of the $\hat{g}_{ii}$.

As before, the condition that F should be diagonal is easily relaxed. All that is necessary is to replace $d_i(s)$ (resp. $\delta_i(s)$) in (ii), but not in (i), by

$$d_i(s) = \sum_{\substack{j=1 \\ j \neq i}}^{m} |f_{ij}(s) + \hat{g}_{ij}(s)| \tag{6.32}$$

$$\delta_i(s) = \sum_{\substack{j=1 \\ j \neq i}}^{m} |f_{ji}(s) + \hat{g}_{ji}(s)| \tag{6.33}$$

However, the most important case is where $F = \operatorname{diag}(k_i)$. Equation (6.31) is then the same as saying that $\hat{\Gamma}_{oi}$ encircles the point $(-k_i, 0)$ as often as it encircles the origin, $i = 1, 2, \ldots, m$. This is an easy criterion to apply because only one curve need be drawn for each i.

Example 6.2 Let a system having least order give rise to

$$G(s) = \begin{pmatrix} \dfrac{4(s+1)}{(2s+1)(2s+3)} & \dfrac{-2}{(2s+1)(2s+3)} \\ \dfrac{-2}{(2s+1)(2s+3)} & \dfrac{4(s+1)}{(2s+1)(2s+3)} \end{pmatrix} \tag{6.34}$$

Because the underlying system has least order, it has $\{\eta_i\} = \{\alpha_i\}$ by Chapter 2,

Section 5, and from (6.34) each α_i is in the open left half-plane, so that the system is asymptotically stable. From (6.34),

$$\hat{G}(s) = \begin{pmatrix} s+1 & \frac{1}{2} \\ \frac{1}{2} & s+1 \end{pmatrix} \tag{6.35}$$

and condition (i) of Theorem (6.4) is

$$|s+1| - \tfrac{1}{2} > 0, \quad s \text{ on } D \tag{6.36}$$

which is easily seen to be satisfied for any finite R however large. If $F =$ diag (k_1, k_2), condition (ii) of the theorem is

$$|s+1+k_i| - \tfrac{1}{2} > 0, \quad s \text{ on } D \tag{6.37}$$

which is true for any $k_i > 0$. Consequently the closed-loop system is asymptotically stable if the inverse Nyquist diagrams corresponding to $s + 1 + k_1$, $s + 1 + k_2$ indicate asymptotic stability in the normal way. This they do for all $k_1 > 0$, $k_2 > 0$, proving that the closed-loop system is asymptotically stable for all $k_1, k_2 > 0$. An alternative procedure is to assimilate the gain into the forward path by putting $\hat{k}_i = k_i^{-1}$ and considering

$$\hat{H} = \begin{pmatrix} 1+\hat{k}_1(s+1) & \frac{1}{2}\hat{k}_1 \\ \frac{1}{2}\hat{k}_2 & 1+\hat{k}_2(s+1) \end{pmatrix} \tag{6.38}$$

Condition (i) in Theorem 6.4 is certainly fulfilled if G is diagonal. The connection with diagonal matrices is reinforced by the following results.

Theorem 6.5 Let the $m \times m$ matrix Q over the field of complex numbers satisfy $\sum_{k=1,k\neq i}^{m} |q_{ik}| \leqslant \theta_i |q_{ii}|$ (resp $\sum_{k=1,k\neq i}^{m} |q_{ki}| \leqslant \theta_i |q_{ii}|$) for $i = 1, 2, \ldots, m$, where $0 \leqslant \theta_i < 1$. Then Q has an inverse $\hat{Q}$ satisfying

$$|\hat{q}_{ji}| \leqslant \theta_j |\hat{q}_{ii}| \text{ (resp } |\hat{q}_{ij}| \leqslant \theta_j |\hat{q}_{ii}|)$$

for $i = 1, 2, \ldots, m$ and $j = 1, 2, \ldots, i-1, i+1, \ldots, m$.

Proof By Gershgorin's theorem Q has no zero eigenvalue and so is non-singular. Then we have

$$\sum_{k=1}^{m} q_{jk}\hat{q}_{ki} = 0, \quad j \neq i \tag{6.39}$$

whence

$$\hat{q}_{ji} + \sum_{\substack{k=1 \\ k\neq j}}^{m} \frac{q_{jk}}{q_{jj}} \hat{q}_{ki} = 0 \tag{6.40}$$

$$|\hat{q}_{ji}| \leqslant \left\{\max_{k \neq j} |\hat{q}_{ki}|\right\} \sum_{\substack{k=1 \\ k \neq j}}^{m} |q_{jk}|/|q_{jj}| \tag{6.41}$$

$$\leqslant \left\{\max_{k} |\hat{q}_{ki}|\right\} \theta_j \tag{6.42}$$

$$< \max_{k} |\hat{q}_{ki}|, \quad j = 1, 2, \ldots, i-1, i+1, \ldots, m \tag{6.43}$$

But this last inequality implies

$$\max_{k} |\hat{q}_{ki}| = |\hat{q}_{ii}| \tag{6.44}$$

when (6.42) gives the desired result. The alternative result is proved in the same way.

EXERCISE 6.1 Show by means of an example that if $m > 2$ the condition

$$\sum_{\substack{k=1 \\ k \neq i}}^{m} |q_{ik}| < |q_{ii}| \qquad \text{implies} \qquad |\hat{q}_{ji}| < |\hat{q}_{ii}|, \quad j \neq i,$$

but does not imply the stronger result

$$\sum_{\substack{j=1 \\ j \neq i}}^{m} |\hat{q}_{ji}| < |\hat{q}_{ii}|$$

[Start from $Q = \begin{pmatrix} 1 & 0 & 0 \\ \alpha & 1 & \beta \\ \alpha & \beta & 1 \end{pmatrix}$].

Theorem 6.6 Let Q satisfy the conditions of Theorem 6.5, and let

$$\phi_i = \max_{k \neq i} \theta_k.$$

Then $|q_{ii} - \hat{q}_{ii}^{-1}| < \theta_i \phi_i |q_{ii}|, \quad i = 1, 2, \ldots, m.$

Proof We have

$$\sum_{k=1}^{m} q_{ik}\hat{q}_{ki} = 1 \tag{6.45}$$

$$q_{ii} + \sum_{\substack{k=1 \\ k \neq 1}}^{m} q_{ik} \frac{\hat{q}_{ki}}{\hat{q}_{ii}} = \hat{q}_{ii}^{-1} \tag{6.46}$$

$$|q_{ii} - \hat{q}_{ii}^{-1}| \leqslant \left\{\sum_{\substack{k=1 \\ k \neq 1}}^{m} |q_{ik}|\right\}\left\{\max_{k \neq i} \left|\frac{\hat{q}_{ki}}{\hat{q}_{ii}}\right|\right\} \tag{6.47}$$

$$\leqslant \theta_i |q_{ii}| \phi_i \tag{6.48}$$

which is the desired result. If the alternative condition holds, the proof follows in a similar way from

$$\sum_{k=1}^{m} \hat{q}_{ik} q_{ki} = 1 \tag{6.49}$$

Theorem 6.7 Let $\hat{G}$ satisfy the following strengthened form of condition (i) in Theorem 6.4: $d_i(s) \leqslant \theta_i(s)|\hat{g}_{ii}(s)|$ (resp $\delta_i(s) \leqslant \theta_i(s)|\hat{g}_{ii}(s)|$) for $i = 1, 2, \ldots, m$ and for all s on D, where $0 \leqslant \theta_i(s) < 1$ and $\max_{k \neq i} \theta_k(s) = \phi_i(s)$. Then

$$|\hat{g}_{ii}(s) - g_{ii}^{-1}(s)| \leqslant \theta_i(s)\phi_i(s)|\hat{g}_{ii}(s)| < |\hat{g}_{ii}(s)|$$

for $i = 1, 2, \ldots, m$ and for all s on D.

Proof Use Theorem 6.6 with $Q = \hat{G}(s)$ for any given s on D.

Corollary 1 Let $\hat{G}$ satisfy condition (i) of Theorem 6.4. If g_{ii} has z_i zeros and p_i poles inside D, and $\hat{g}_{ii}$ has $\hat{z}_i$ zeros and $\hat{p}_i$ poles inside D, then

$$z_i - p_i = -(\hat{z}_i - \hat{p}_i), \quad i = 1, 2, \ldots, m \tag{6.50}$$

Proof This follows from Theorem 6.7, but it is instructive to give a more direct proof. If $\hat{G}$ satisfies condition (i) of Theorem 6.4, then *a fortiori* any submatrix obtained from $\hat{G}$ by deleting rows $i_1, i_2, \ldots, i_k$ and columns $i_1, i_2, \ldots, i_k$, has the same property. Now

$$g_{ii} = \frac{\hat{G}_{ii}}{|\hat{G}|} \tag{6.51}$$

and by Theorem 6.2 applied to $\hat{G}_{ii}$ and to $|\hat{G}|$,

$$z_i - p_i = \sum_{\substack{j=1 \\ j \neq i}}^{m} (\hat{z}_j - \hat{p}_j) - \sum_{j=1}^{m} (\hat{z}_j - \hat{p}_j) \tag{6.52}$$

$$= -(\hat{z}_i - \hat{p}_i) \tag{6.53}$$

Corollary 2 If no pole of G, and no zero of $|G|$, lies inside D, then

$$z_i = \hat{z}_i = p_i = \hat{p}_i = 0, \quad i = 1, 2, \ldots, m \tag{6.54}$$

Proof No pole of G, and therefore no pole of any g_{ii} lies inside D, whence $p_i = 0$, $i = 1, 2, \ldots, m$. Also

$$\hat{g}_{ii} = \frac{G_{ii}}{|G|} \tag{6.55}$$

which also has no pole inside D, showing that $\hat{p}_i = 0$, $i = 1, 2, \ldots, m$. Then (6.50) gives $z_i = -\hat{z}_i$, whence the result.

Theorem 6.7 can be applied to $\hat{H}$ if this satisfies a suitably strengthened form of condition (ii) in Theorem 6.4. The effect is to locate the inverse transfer function $h_{ii}^{-1}(s)$ seen between input i and output i within a band centered on $\hat{h}_{ii}(s) = \hat{g}_{ii}(s) + f_{ii}(s)$.

6.2 Systems described by difference equations

Let feedback be applied to the system

$$\tilde{T}(z)\tilde{\xi} = \tilde{U}(z)\tilde{u} \tag{6.56}$$

$$\tilde{y} = \tilde{V}(z)\tilde{\xi} + \tilde{W}(z)\tilde{u} \tag{6.57}$$

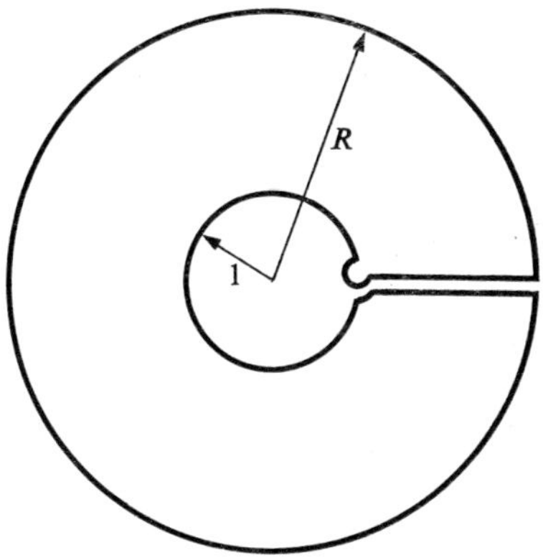

FIG. 1 The unit circle is shown indented for a pole at $z = 1$

in which $l = m$, according to the equation,

$$\tilde{u} = \tilde{v} - F\tilde{y} \tag{6.58}$$

where F is a constant matrix. Apart from the fact that asymptotic stability now requires that the appropriate roots be inside the unit circle, rather than in the open left half-plane, the situation remains very much as in Section 6.1.

The contour D is now replaced by a contour C, as in Fig. 1. This consists first of the unit circle, indented towards the centre to avoid any pole or zero of $|\hat{\hat{G}}|$, $|\hat{\hat{H}}|$, $\hat{\hat{g}}_{ij}$, or $\hat{\hat{h}}_{ij}$ on this circle. The contour C is completed by two adjacent lines and a circle of radius R. The two lines and the large circle are chosen so that they do not intersect any pole or zero of $|\hat{\hat{G}}|$, $|\hat{\hat{H}}|$, $\hat{\hat{g}}_{ij}$, or $\hat{\hat{h}}_{ij}$. Finally, R is chosen large enough to ensure that every finite pole or zero of $|\hat{\hat{G}}|$, $|\hat{\hat{H}}|$, g_{ij}, and $\hat{\hat{h}}_{ij}$ lies inside the large circle. In this specification of C, reference to $\hat{\tilde{g}}_{ij}$ and $\hat{\tilde{h}}_{ij}$ can be omitted except in results which appeal to Gershgorin's theorem.

Let the contour C be mapped by $|\hat{G}|$ into Γ_0, by $|\hat{\tilde{G}}|$ into $\hat{\Gamma}_0$, by $|\tilde{H}|$ into Γ_c, by $|\hat{\tilde{H}}|$ into $\hat{\Gamma}_c$, by $\hat{\tilde{g}}_{ii}$ into $\hat{\Gamma}_{0i}$, and by $\hat{\tilde{h}}_{ii}$ into $\hat{\Gamma}_{ci}$. Let $\hat{\Gamma}_{0i}$ encircle the origin N_{0i} times, and let $\hat{\Gamma}_{ci}$ encircle the origin N_{ci} times.

The results corresponding to Theorems 6.1, 6.3, and 6.4 are now easy to establish, and are left as exercises for the reader.

EXERCISE 6.2 Let (6.56) be asymptotically stable. Show that the closed-loop system is asymptotically stable if and only if Γ_c, Γ_0 encircle the origin the same number of times.

EXERCISE 6.3 Let (6.56) be asymptotically stable. Show that the closed-loop system is asymptotically stable if and only if $\hat{\Gamma}_c$, $\hat{\Gamma}_0$ encircle the origin the same number of times.

EXERCISE 6.4 Let (6.56) be asymptotically stable, let $|\tilde{G}|$ have no zero on or outside the unit circle, and let F be diagonal. For all z on C and $i = 1, 2, \ldots, m$, let

$$|f_{ii} + \hat{\tilde{g}}_{ii}(z)| - \sum_{\substack{j=1\\ j\neq i}}^{m} |\hat{\tilde{g}}_{ij}(z)| > 0 \tag{6.59}$$

or alternatively for all z on C and $i = 1, 2, \ldots, m$, let

$$|f_{ii} + \hat{\tilde{g}}_{ii}(z)| - \sum_{\substack{j=1\\ j\neq i}}^{m} |\hat{\tilde{g}}_{ji}(z)| > 0 \tag{6.60}$$

Show that the closed-loop system is asymptotically stable if and only if

$$\sum_{i=1}^{m} N_{ci} = 0 \tag{6.61}$$

EXERCISE 6.5 Let (6.45) be asymptotically stable, and let F be diagonal. For all z on C and for $i = 1, 2, \ldots, m$ let

(i) $|\hat{\tilde{g}}_{ii}(z)| - \sum_{\substack{j=1\\ j\neq i}}^{m} |\hat{\tilde{g}}_{ij}(z)| > 0$ (resp. $|\hat{\tilde{g}}_{ii}(z)| - \sum_{\substack{j=1\\ j\neq i}}^{m} |\hat{\tilde{g}}_{ji}(z)| > 0$)

(ii) $|f_{ii} + \hat{\tilde{g}}_{ii}(z)| - \sum_{\substack{j=1\\ j\neq i}}^{m} |\hat{\tilde{g}}_{ij}(z)| > 0$ (resp. $|f_{ii} + \hat{\tilde{g}}_{ii}(z)| - \sum_{\substack{j=1\\ j\neq i}}^{m} |\hat{\tilde{g}}_{ji}(z)| > 0$)

Show that the closed-loop system is asymptotically stable if and only if

$$\sum_{i=1}^{m} N_{ci} = \sum_{i=1}^{m} N_{0i} \tag{6.62}$$

7 Structure of multivariable systems

When a multivariable control system has to be designed, the possible choice of structure seems at first sight to be embarrassingly wide. For example, if an $m \times m$ transfer function matrix $G(s)$ is given, we may wish to find an $m \times m$ compensator $K(s)$ such that the closed-loop system, with GK in the forward path and $F = I_m$ in the feedback path, has desired behaviour. We shall wish to have $K(s)$ a rational matrix, but this leaves a wide choice for its form.

As a practical matter, we shall usually restrict K to have all its poles in the open left half-plane. Also, in view of Theorem 2.7, we shall wish $|K|$ to have all its zeros in the open left half-plane. Subject to these conditions a number of results can be proved about the structure of $K(s)$, and about the effect which K can have upon G.

Theorem 7.1 Let $K(s)$ be a non-singular $m \times m$ rational matrix. Let all the poles of $K(s)$, and all the zeros of $|K(s)|$, lie in the open left half-plane. Then $K(s)$ can be written as a product $K_a K_b(s) K_c(s)$, where

(i) K_a is a permutation matrix,
(ii) $K_b(s)$ is the product of elementary matrices $K_b^{(k)}$,

$$K_b^{(k)}(s) = \begin{pmatrix} 1 & 0 & \dots & 0 & \dots & 0 & \dots & 0 \\ 0 & 1 & \dots & 0 & \dots & 0 & \dots & 0 \\ \vdots & \vdots & & \vdots & & \vdots & & \vdots \\ 0 & 0 & \dots & 1 & \dots & \alpha_{ij}^{(k)}(s) & \dots & 0 \\ \vdots & \vdots & & \vdots & & \vdots & & \vdots \\ 0 & 0 & \dots & 0 & \dots & 0 & \dots & 1 \end{pmatrix} \tag{7.1}$$

Here $\alpha_{ij}^{(k)}(s)$ can occur in any position with $j \neq i$, and is a rational function having all its poles in the open left half-plane.
(iii) $K_c(s)$ is a non-singular diagonal matrix. All the poles and zeros of the non-zero (diagonal) elements of $K_c(s)$ lie in the open left half-plane.

Proof Let $p_i(s)$ be the least common denominator of the ith column of $K(s)$: clearly each $p_i(s)$ has its zeros in the open left half-plane. In the polynomial matrix $K(s)\operatorname{diag}(p_i(s))$, let $D_i(s)$ be the greatest common divisor of all $i \times i$ minors formed from the first i columns. From Laplace's expansion of a determinant it follows that $D_i(s)$ divides $D_{i+1}(s)$, $i = 1, 2, \dots, m-1$. Because a zero of $D_m(s)$ is a zero of $|K(s)|$ or of a $p_i(s)$, it follows that each $D_i(s)$ has all its zeros in the open left half-plane.

Multiply the first column of $K(s)$ by $p_1(s)/D_1(s)$. The resulting column will have polynomial entries with no common divisor. By a procedure used in Theorem 1.6 of Chapter 1, this column can therefore be reduced to the form

$(a_1, 0, 0, \ldots, 0)^T$, where a_1 is a non-zero constant. The operations needed in this procedure are of the following types:

(a) transpose two rows;
(b) add to row i a multiple by $\alpha_{ij}^{(k)}(s)$ of row j, where $\alpha_{ij}^{(k)}(s)$ is a polynomial in s.

These operations do not change the greatest common divisor $D_2(s)$ of all 2×2 minors formed from the first two columns. It follows that if the second column of the matrix obtained in the above way is multiplied by $p_2(s)D_1(s)/D_2(s)$, the elements in positions $(2, 2), (3, 2), \ldots, (m, 2)$ are then polynomials with no common divisor. As before, the $(m - 1)$-vector formed by these elements can be reduced to the form $(a_2, 0, 0, \ldots, 0)^T$ by row operations of types (a) and (b), applied to the last $m - 1$ rows. The element in position $(1, 2)$ is a rational function with its poles in the open left half-plane. Subtract a suitable multiple (by a rational function) of row 2 from row 1 and so reduce the element $(1, 2)$ to zero. This does not affect the element $(1, 1)$, so that the matrix now has the form

$$\left(\begin{array}{cc|c} a_1 & 0 & \\ 0 & a_2 & T_1(s) \\ \hline \multicolumn{2}{c|}{0} & T_2(s) \end{array}\right) \tag{7.2}$$

where T_1 is $2 \times (m - 2)$ and T_2 is $(m - 2) \times (m - 2)$.

Proceeding in this way we can reduce the matrix to the form diag (a_i). Write $D_0(s) = 1$, and put

$$K_c(s) = \operatorname{diag}\left(\frac{a_i D_i(s)}{p_i(s)D_{i-1}(s)}\right) \tag{7.3}$$

The above argument shows that

$$K_d(s)K(s) = K_c(s) \tag{7.4}$$

where $K_d(s)$ is generated by operations of types (a) and (b), but with $\alpha_{ij}(s)$ in (b) now either a polynomial, or a rational function with all its poles in the open left half-plane.

Each of these two types of operation has an inverse satisfying the same description. Moreover, if K_e represents a transposition of two rows, and K_f a matrix of the type (7.1), the matrix $K_e^{-1}K_fK_e$ has every entry on the principal diagonal 1, and every other entry zero except for one entry $\alpha_{ij}^{(k)}(s)$ equal to the off-diagonal entry in K_f, but (in general) in a different position. That is, $K_fK_e = K_eK_f'$, where K_f' is another matrix of the type (7.1). It follows that $K_d^{-1}(s)$ can be written $K_aK_b(s)$, where K_a and K_b have the properties stated in the theorem. By (7.3) and (7.4), this completes the proof.

Example 7.1 Consider

$$K(s) = \begin{pmatrix} \dfrac{1-s}{(s+1)^2} & \dfrac{-s}{s+1} \\ \dfrac{1}{s+1} & 1 \end{pmatrix} \tag{7.5}$$

By the procedure given in the proof of Theorem 7.1, K can be put in the form,

$$\begin{aligned} K(s) &= K_b(s)K_c(s) \\ &= \left[\begin{pmatrix} 1 & -1 \\ 0 & 1 \end{pmatrix}\begin{pmatrix} 1 & 0 \\ \dfrac{s+1}{2} & 1 \end{pmatrix}\begin{pmatrix} 1 & \dfrac{2}{s+1} \\ 0 & 1 \end{pmatrix}\right]\begin{pmatrix} \dfrac{2}{(s+1)^2} & 0 \\ 0 & \frac{1}{2} \end{pmatrix} \end{aligned} \tag{7.6}$$

Notice that if $K(s)$ is regarded as a matrix over the field of rational functions the ordinary process of Gauss reduction leads to

$$\begin{aligned} K(s) &= K_b'(s)K_c'(s) \\ &= \left[\begin{pmatrix} 1 & 0 \\ \dfrac{s+1}{1-s} & 1 \end{pmatrix}\begin{pmatrix} 1 & \dfrac{s(s-1)}{s+1} \\ 0 & 1 \end{pmatrix}\right]\begin{pmatrix} \dfrac{1-s}{(s+1)^2} & 0 \\ 0 & \dfrac{1}{1-s} \end{pmatrix} \end{aligned} \tag{7.7}$$

but here K_b' and K_c' both have poles in the right half-plane.

Exercise 7.1 Let a plant have the $m \times m$ transfer function matrix $G(s)$, and let an $m \times m$ compensator $K(s)$ have been designed by some suitable technique. Let feedback be applied to the compensated plant $G(s)K(s)$ in such a way that the closed-loop transfer function matrix is $H = GK[I_m + GK]^{-1}$. Show that if $K(s)$ satisfies the conditions of Theorem 7.1, the closed-loop system can be regarded as having

(i) a set of m single-loop controllers; followed by
(ii) a compensator $K_b(s)$, with $|K_b(s)| = 1$, which recombines the parallel paths through the system; followed by
(iii) a compensator K_a which has the effect of renumbering the inputs.

Show also that the order in which $K_b(s)$ and K_a occur can be inverted if desired, with a consequent change in $K_b(s)$.

Theorem 7.2 Let $G(s)$ be a non-singular $m \times m$ rational matrix. Let all the poles of $G(s)$, and all the zeros of $|G(s)|$, lie in the open left half-plane. Then K_a and $K_b(s)$ can be found, defined as in Theorem 6.1, such that $G(s)K_aK_b(s)$ is non-singular, diagonal, and has all the poles and zeros of its non-zero (diagonal) elements in the open left half-plane.

Proof Interchanging the roles of rows and columns in the proof of Theorem 7.1, it follows that $G(s)$ can be written $G(s) = G_c(s)G_b(s)G_a$. Here G_a is a permutation matrix, $G_b(s)$ is a product of matrices having the form given in (7.1), and $G_c(s)$ is non-singular, diagonal, and with all the poles and zeros of its non-zero (diagonal) elements in the open left half-plane. The choice

$$K_a = G_a^{-1} \tag{7.8}$$

$$K_b(s) = G_b^{-1}(s) \tag{7.9}$$

accomplishes the desired result.

Theorem 7.3 Let $G(s)$ satisfy the conditions of Theorem 7.2 except that $|G(s)|$ has one or more zeros in the closed right half-plane. Then K_a and $K_b(s)$ can be found, defined as in Theorem 7.1, such that $G(s)K_aK_b(s)$ is triangular (upper or lower as desired). In general $G(s)K_aK_b(s)$ cannot be made diagonal.

Proof Let $p_i(s)$ be the least common denominator of the ith row of $G(s)$: each $p_i(s)$ has its zeros in the open left half-plane. In the polynomial matrix diag $(p_i(s))G(s)$, let $D_i(s)$ be the greatest common divisor of all $i \times i$ minors formed from the first i rows, and put $D_0(s) = 1$. Let

$$D_i(s) = D_{il}(s)D_{ir}(s) \tag{7.10}$$

where $D_{il}(s)$ has all its zeros in the open left half-plane and $D_{ir}(s)$ has all its zeros in the closed right half-plane. It follows from Laplace's expansion of a determinant that $D_{il}(s)$ (resp. $D_{ir}(s)$) divides $D_{i+1,l}(s)$ (resp. $D_{i+1,r}(s)$), $i = 1, 2, \ldots, m-1$.

The first row of diag $(p_i(s))G(s)$ has polynomial entries with greatest common divisor $D_1(s)$. By the procedure described in Theorem 1.6 of Chapter 1, this row can therefore be brought to the form $(a_1D_1(s), 0, 0, \ldots, 0)$, where a_1 is a non-zero constant, by operations of the following types:

(a) Transpose two columns.
(b) Add to column j a multiple by $\alpha_{ij}^{(k)}(s)$ of column i, where $\alpha_{ij}^{(k)}(s)$ is a polynomial in s.

These operations do not change the greatest common divisor $D_2(s)$ of the 2×2 minors formed from the first two rows: the elements in positions (2, 2), (2, 3), ..., (2, m) are therefore divisible by $D_2(s)/D_1(s)$. By operations of types (a) and (b), the last $m-1$ elements in the second row can be brought to the form $(a_2D_2(s)/D_1(s), 0, 0, \ldots, 0)$ where a_2 is a non-zero constant. Proceeding in this way we generate a lower triangular polynomial matrix having as the elements on its principal diagonal

$$a_i\frac{D_i(s)}{D_{i-1}(s)} = a_i\frac{D_{il}(s)}{D_{i-1,l}(s)}\frac{D_{ir}(s)}{D_{i-1,r}(s)}, \quad i = 1, 2, \ldots, m \tag{7.11}$$

If the element in position $(m, m-1)$ is divisible by $D_{mr}(s)/D_{m-1,r}(s)$, subtract such a multiple of the element in position (m, m) [by a rational function having denominator $D_{ml}(s)/D_{m-1,l}(s)$] that it is reduced to zero. Proceed similarly with elements in positions $(m, m-2)$, $(m, m-3), \ldots,$ $(m, 1)$, $(m-1, m-2)$, $(m-1, m-3), \ldots$. Stop when an element is reached in row i which is not divisible by $D_{ir}(s)/D_{i-1,r}(s)$, or when the matrix has been made diagonal. Call the resulting lower triangular polynomial matrix $T_1(s) = (\tau_{ij}(s))$. [The manipulation in this paragraph is needed for the purposes of Theorem 7.4, and may be omitted if the only requirement is to bring G to triangular form.]

As in the proof of Theorem 6.1 it follows that

$$G(s) = \operatorname{diag}\left(\frac{1}{p_i(s)}\right) T_1(s) G_b(s) G_a \tag{7.12}$$

Here G_a is a permutation matrix and $G_b(s)$ can be written as a product of matrices such as (7.1). Setting $K_a = G_a^{-1}$, $K_b(s) = G_b^{-1}(s)$ it follows that

$$G(s) K_a K_b(s) = T(s) \tag{7.13}$$

where $T(s)$ is a lower triangular matrix. With minor changes in the proof an upper triangular matrix can be generated.

That reduction to diagonal form is not always possible with K_a and $K_b(s)$ as defined in Theorem 7.1 is shown by the example

$$G(s) = \begin{pmatrix} \dfrac{1-s}{(s+1)^2} & \dfrac{1}{2s+1} \\ 0 & \dfrac{1}{s+1} \end{pmatrix} \tag{7.14}$$

We attempt to find $K(s)$ such that

$$G(s)K(s) = \begin{pmatrix} \dfrac{1-s}{(s+1)^2} & \dfrac{1}{2s+1} \\ 0 & \dfrac{1}{s+1} \end{pmatrix} \begin{pmatrix} k_{11}(s) & k_{12}(s) \\ k_{21}(s) & k_{22}(s) \end{pmatrix} \tag{7.15}$$

is diagonal. It then follows that $k_{21}(s) = 0$, and

$$\frac{1-s}{(s+1)^2} k_{12}(s) + \frac{1}{2s+1} k_{22}(s) = 0 \tag{7.16}$$

Consequently,

$$|K(s)| = k_{11}(s) k_{22}(s) = -k_{11}(s) k_{12}(s) \frac{(2s+1)(1-s)}{(s+1)^2} \tag{7.17}$$

No choice of $k_{11}(s)$ and $k_{12}(s)$ without poles in the closed right half-plane can avoid a factor $s - 1$ in $|K(s)|$. But with the definitions of K_a and $K_b(s)$, $|K_a| = \pm 1$, $|K_b(s)| = 1$, so that $K(s)$ cannot be represented by $K_a K_b(s)$.

Theorem 7.4 Let $G(s)$ be as in Theorem 7.3. Find K_a and $K_b(s)$ as in that theorem so that (7.13) is true. Then $T(s)$ may be diagonal. If not, let $K_e(s)$ be any non-singular rational matrix having all its poles in the open left half-plane and such that $G(s)K_e(s)$ is diagonal. Then $|K_e(s)|$ has at least one zero in the closed right half-plane.

Proof Assume $T(s)$ to be lower triangular. We have

$$G(s)K_aK_b(s)[K_b^{-1}(s)K_a^{-1}K_e(s)] = T(s)K_f(s) = D(s) \tag{7.18}$$

where $D(s) = \big(d_{ij}(s)\big)$ is diagonal and $K_f(s) = \big(\kappa_{ij}(s)\big)$. Hence

$$K_f(s) = T^{-1}(s)D(s) \tag{7.19}$$

and so $K_f(s)$ is lower triangular. It is also non-singular, and has all its poles in the open left half-plane.

If $T(s)$ is not diagonal, there is a $t_{ii}(s)$ having a zero in the closed right half-plane, and a $t_{ij}(s) \neq 0$ with $i > j$ such that

$$t_{i,j+1}(s) = t_{i,j+2}(s) = \cdots = t_{i,i-1}(s) = 0 \tag{7.20}$$

Then in (7.18),

$$0 = d_{ij}(s) = t_{ij}\kappa_{jj}(s) + t_{ii}(s)\kappa_{ij}(s) \tag{7.21}$$

and so

$$\kappa_{jj}(s) = -\frac{\kappa_{ij}(s)t_{ii}(s)}{t_{ij}(s)} \tag{7.22}$$

From the proof of Theorem 7.3,

$$t_{ii}(s) = \frac{a_i}{p_i(s)}\left[\frac{D_{il}(s)}{D_{i-1,l}(s)}\right]\left[\frac{D_{ir}(s)}{D_{i-1,r}(s)}\right] \tag{7.23}$$

and

$$t_{ij}(s) = \frac{\tau_{ij}(s)}{p_i(s)} \tag{7.24}$$

where $\tau_{ij}(s)$ is an element of $T_1(s)$. Because the polynomial $D_{ir}(s)/D_{i-1,r}(s)$ has all its zeros in the closed right half-plane and does not divide $\tau_{ij}(s)$, it follows from (7.22) and the properties of $\kappa_{ij}(s)$ and the $D_{il}(s)$ that $\kappa_{jj}(s)$ has a zero in the closed right half-plane. Consequently $|K_f(s)|$ has a zero there, and by (7.18) and the properties of K_a and $K_b(s)$, so does $|K_e(s)|$. The proof holds with minor changes if T is upper triangular, and it is exemplified by equation (7.17).

In this section we have talked of 'the zeros of $|K(s)|$ lying in the closed right half-plane' for example, rather than 'the zeros of $K(s)$ lying in the closed right half-plane'. The reason why this is possible is that $K(s)$ was assumed to have all its poles in the open left half-plane. Therefore in the McMillan form of $K(s)$, no zero of an $\varepsilon_i(s)$ lying in the closed right half-plane can cancel a zero of any $\psi_j(s)$ when $|K(s)|$ is formed.

If a system is described by difference equations, similar results can be stated for its z-transfer function $\tilde{G}(z)$ and any compensator $\tilde{K}(z)$. Throughout the theorems, the open left half-plane is replaced by the interior of the unit circle, and the closed right half-plane by the perimeter and the exterior of this circle.

8 The definition of state

This completes our development of that part of the theory dealing with constant linear systems. We may now return to the definition of the order of a system which was adopted in Section 2, Chapter 2. As the order is the dimension of the state vector, we are really concerned with the definition of state. Intuitively, we regard the state as summarizing all information about the past which we need for calculating those aspects of future behaviour in which we are interested. This is not purely a matter of mathematics: the definition we adopt can be regarded in part as a statement of our interest.

This may be made clearer by an example. Suppose that there are two objects, one obeying the equation

$$\dot{x} = A_1x_1 + B_1u_1 \tag{8.1}$$

and the other obeying

$$\dot{x}_2 = A_2x_2 \tag{8.2}$$

The first is situated in Birmingham, England, and the second in Birmingham, Alabama, and there is no connection of any kind between them. We should agree that the state of the first object is x_1. No doubt we should remain of this opinion even if a mathematician with a sense of humour had carried out a preliminary transformation

$$\xi = H\begin{pmatrix} x_1 \\ x_2 \end{pmatrix} \tag{8.3}$$

and presented us with the equations

$$\dot{\xi} = A\xi + Bu \tag{8.4}$$

where

$$A = H\begin{pmatrix} A_1 & 0 \\ 0 & A_2 \end{pmatrix}H^{-1}, \qquad B = H\begin{pmatrix} B_1 \\ 0 \end{pmatrix} \tag{8.5}$$

If the two objects described by (8.1) and (8.2) were in the same box, should we continue to regard x_1 separately as a state for the first object? Possibly we might, and possibly we might not. If we obtained the equations for the two objects simultaneously in the form (8.4), we should almost certainly say that the state was ξ. Then by a suitable transformation we could exhibit x_2 as the uncontrollable (p.s.) part of the state.

It would be open to us, however, to define the state, even for (8.4), as only the controllable (p.s.) part. This might be appropriate if we were concerned only with the behaviour starting from zero initial conditions. The convention which has grown up of defining the state to include the uncontrollable (p.s.) part if any is, however, a good one. If one of the uncontrollable modes is unstable, we may still need to pay attention to it. Our interest in this case is not strictly limited to zero initial conditions, but extends to initial conditions in some neighbourhood of the origin.

A similar point arises in Chapter 2. The state there was identified with the independent coefficients in the complementary function. It would have been equally possible to identify the state with the initial conditions used in applying the Laplace transformation. Then the system matrix in equation (4.15) of Chapter 2, for example, would not be regarded as having least order, and order would not remain invariant under strict system equivalence. Such a procedure would be quite consistent with the usual definition of order, and the objections to it must be on grounds of convenience rather than logic.

These questions have more than an academic interest. If the equations of a nonlinear plant are linearized, it is quite possible to obtain derivatives of the control variable u in the equations. Suppose we have obtained

$$\frac{d}{dt}\begin{pmatrix} x_1 \\ x_2 \end{pmatrix} = \begin{pmatrix} 0 & 1 \\ 0 & 1 \end{pmatrix}\begin{pmatrix} x_1 \\ x_2 \end{pmatrix} + \begin{pmatrix} 0 \\ 1 \end{pmatrix}\dot{u} = Ax + b\dot{u} \tag{8.6}$$

Are we to say that this system is controllable (p.s.) because

$$(b, Ab) = \begin{pmatrix} 0 & 1 \\ 1 & 1 \end{pmatrix} \tag{8.7}$$

has rank 2? If we write (8.6) in the state-space form

$$\frac{d}{dt}\left[\begin{pmatrix} x_1 \\ x_2 \end{pmatrix} - \begin{pmatrix} 0 \\ 1 \end{pmatrix}u\right] = \begin{pmatrix} 0 & 1 \\ 0 & 1 \end{pmatrix}\left[\begin{pmatrix} x_1 \\ x_2 \end{pmatrix} - \begin{pmatrix} 0 \\ 1 \end{pmatrix}u\right] + \begin{pmatrix} 1 \\ 1 \end{pmatrix}u \tag{8.8}$$

we obtain instead of (8.7)

$$\begin{pmatrix} 1 & 1 \\ 1 & 1 \end{pmatrix} \tag{8.9}$$

which has rank 1.

The apparent paradox arises because in writing (8.7) we are tacitly putting

$$\dot{u} = v \tag{8.10}$$

and treating v as the input. This changes the transfer function and also the order of the system: (8.6) gives, for $y = x$,

$$G(s) = \frac{1}{s-1}\begin{pmatrix} 1 \\ s \end{pmatrix} \tag{8.11}$$

and the system has second order. The system

$$\left.\begin{aligned} \dot{u} &= v \\ \begin{pmatrix} \dot{x}_1 \\ \dot{x}_2 \end{pmatrix} &= \begin{pmatrix} 0 & 1 \\ 0 & 1 \end{pmatrix}\begin{pmatrix} x_1 \\ x_2 \end{pmatrix} + \begin{pmatrix} 0 \\ 1 \end{pmatrix} v \end{aligned}\right\} \tag{8.12}$$

on the other hand has order 3 and gives

$$G(s) = \frac{1}{s(s-1)}\begin{pmatrix} 1 \\ s \end{pmatrix} \tag{8.13}$$

We cannot say that one of these procedures is right and the other wrong: either might be appropriate in given circumstances. However, in replacing (8.6) by (8.12) we are changing our mind about what we regard as important in a much deeper sense than when we replace (8.6) by (8.8). Therefore the transformation from (8.6) to (8.8) is likely to have much wider application. It is in fact a transformation of strict system equivalence.

The definition of system order and the transformations introduced in Chapter 2 are guided by two considerations:

(i) Manipulation of the equations should not change G.
(ii) Manipulation of the equations should leave the system order unchanged.

Beyond this, all the manipulations usually employed to bring equations into state-space form should be allowed. When (i) has been used to eliminate transformations such as the one which brings (8.6) to (8.12), the remaining transformations are those permitted by system equivalence. This transformation does not keep system matrices in polynomial form, that is, the form in which they have a one-to-one correspondence with the differential equations. Strict system equivalence is a restriction of system equivalence which preserves the polynomial form. Then, in effect, order was defined so that it remains invariant under this transformation.

The justification for the definitions adopted is to be found in Fig. 1 of Chapter 3. This shows that no conflict will ever arise between the new definitions and the usual definition employed when the equations are in state-space form. From what has been said it will be clear that consistency is all that can be looked for.

Problems

1 Regard the $m \times m$ rational transfer function matrix $G(s)$ as a matrix over the field of rational functions, and write

$$|\mu I_m - G(s)| = \mu^m + a_{m-1}(s)\mu^{m-1} + \cdots + a_0(s)$$

Show that

$$G^m(s) + a_{m-1}(s)G^{m-1}(s) + \cdots + a_0(s)I_m = 0$$

2 Regarding $G(s)$ as in Problem 1, let $M(s)$ be a matrix such that $M^{-1}GM$ is in Jordan form J. Show from equations (1.5) and (1.25) that if $F = I_m$, the closed-loop poles α_i' are zeros of $1 + \mu_j(s) = 0$, $j = 1, 2, \ldots, m$, where the μ_j are the roots of $|\mu I_m - G(s)| = 0$. [Note that the μ_j lie in general in an extension field.]

3 By means of Gershgorin's theorem show that for given s every $\mu_k(s)$ in Problem 2 lies in the union of the disks

$$|g_{ii}(s) - z(s)| \leqslant \sum_{\substack{j=1 \\ j \neq i}}^{m} |g_{ij}(s)|, \quad i = 1, 2, \ldots, m$$

Hence prove a stability theorem similar to Theorem 6.3 but with g_{ii} in place of $\hat{g}_{ii}$ and with the other appropriate changes.

4 Using the result of Problem 3, and the fact that the eigenvalues of G^{-1} are μ_i^{-1}, prove Theorem 6.3.

5 In a closed-loop system, the matrix $I_m + G$ is called the return-difference matrix. Show that its determinant is

$$|I_m + G(s)| = \frac{\prod_{i=1}^{n}(s - \alpha_i')}{\prod_{i=1}^{n}(s - \alpha_i)} = \prod_{j=1}^{m}(1 + \mu_j(s))$$

where the α_i are the open-loop poles, the α_i' are the closed-loop poles, and the μ_j are the roots of $|\mu I_m + G(s)| = 0$. [Compare Problem 2.]

6 Let λ_1 be a real eigenvalue of A, and let b be the corresponding eigenvector of A. Show that the closed-loop system

$$\dot{x} = Ax + bu$$

$$u = -c^{\mathrm{T}}x$$

has an eigenvalue $\lambda_1 - c^{\mathrm{T}}b$.

7 In Problem 6 let A have n linearly independent eigenvectors (for example, let the eigenvalues of A be distinct). If c is orthogonal to every eigenvector of A except b, show that the eigenvectors of $A - bc^{\mathrm{T}}$ are those of A.

8 Show directly that if b is a column vector and $sI_n - A, b$ are relatively (left) prime there exists a row vector c such that

$$|sI_n - A + bc| = \begin{vmatrix} sI_n - A & b \\ -c & 1 \end{vmatrix} = \varphi(s)$$

where $\varphi(s)$ is any monic polynomial of degree n. [Reduce the problem to the consideration of

$$\begin{vmatrix} s & -1 & \cdots & 0 & 0 & 0 \\ 0 & s & \cdots & 0 & 0 & 0 \\ \vdots & \vdots & & \vdots & \vdots & \vdots \\ 0 & 0 & \cdots & s & -1 & 0 \\ a_0 & a_1 & \cdots & a_{n-2} & s + a_{n-1} & 1 \\ c_1 & c_2 & \cdots & c_{n-1} & c_{n-2} & 1 \end{vmatrix}$$

where $c_1, c_2, \ldots, c_n$ are to be chosen.]

9 Let $T(s)$ be an $m \times m$ polynomial matrix with $|T(s)|$ monic of degree n, and let $\phi(s)$ be a monic polynomial of degree n. Show that there exists an $m \times m$ polynomial matrix $V(s)$ such that

(a) T and V are relatively (right) prime;
(b) $VT^{-1} \to 0$ as $s \to \infty$;
(c) $|T(s) + V(s)| = \phi(s)$.

[Use Corollary 2 of Theorem 4.2 with the transformations

$$M\begin{pmatrix} I_{n-m} & 0 & 0 \\ 0 & T & I_m \end{pmatrix}\begin{pmatrix} N & Y \\ 0 & I_m \end{pmatrix} = (sI_n - A \quad B)$$

$$\begin{pmatrix} M^{-1} & 0 \\ X & I_m \end{pmatrix}\begin{pmatrix} sI_n - A & B \\ -C & I_m \end{pmatrix}\begin{pmatrix} N & Y \\ 0 & I_m \end{pmatrix}^{-1} = \begin{pmatrix} I_{n-m} & 0 & 0 \\ 0 & T & I_m \\ 0 & -V & I_m \end{pmatrix}$$

to generate V.]

10 Let $G(s)$ be a proper $m \times m$ rational matrix with $\nu(G) = n$, and let $\phi(s)$ be a monic polynomial of degree n. Show that there does not necessarily exist a polynomial matrix $L(s)$ such that

(a) $L(s)G(s)$ is proper;
(b) $|I_m + L(s)G(s)|$ has the same zeros as $\phi(s)$.

[Give an example for $m = 1$, or alternatively write $G = T^{-1}U$ and show that a procedure similar to that in Problem 9 is not generally successful. Compare Problem 8 of Chapter 2.]

11 Generalize Theorems 6.1, 6.3, and 6.4 to the case where $\phi_0(s)$ in equation (6.4) has zeros in the closed right half-plane.

6 Time-dependent systems

1 Introduction

In previous chapters we have been concerned with systems described by equations such as

$$\dot{x}(t) = Ax(t) + Bu(t) \tag{1.1}$$

$$y(t) = Cx(t) + Du(t) \tag{1.2}$$

where A, B, C, D are constant matrices, respectively $n \times n, n \times l, m \times n, m \times l$. A more general class of systems is obtained by allowing these matrices to be functions of time, so that

$$\dot{x}(t) = A(t)x(t) + B(t)u(t) \tag{1.3}$$

$$y(t) = C(t)x(t) + D(t)u(t) \tag{1.4}$$

Equations such as (1.3) and (1.4) often arise when small perturbations x are considered about a *reference* (or *nominal*) motion of a nonlinear system.

Example 1.1 Let a system obey the nonlinear equations

$$\dot{\xi} = f(\xi, v); \qquad \xi(t_0) = c \tag{1.5}$$

$$\eta = g(\xi, v) \tag{1.6}$$

where ξ is an n-vector, v is an l-vector of inputs, and η is an m-vector of outputs. Let all the first partial derivatives of f and g be continuous and bounded for all ξ and v. Then if v^0 is some given function, and ξ^0, η^0 are the corresponding solutions of (1.5) and (1.6), we obtain for an input $v^0 + u$,

$$\dot{\xi}^0 + \dot{x} = f(\xi^0 + x, v^0 + u); \qquad \xi^0(t_0) + x(t_0) = c \tag{1.7}$$

$$\eta^0 + y = g(\xi^0 + x, v^0 + u) \tag{1.8}$$

where ξ, η have been written $\xi^0 + x, \eta^0 + y$. Expanding the functions by means of their Taylor series we have, to the first order,

$$f_i(\xi^0 + x, v^0 + u) = f_i(\xi^0, v^0) + \sum_{j=1}^{n} \frac{\partial f_i}{\partial \xi_j} x_j + \sum_{j=1}^{l} \frac{\partial f_i}{\partial v_j} u_j \tag{1.9}$$

$$g_i(\xi^0 + x, v^0 + u) = g_i(\xi^0, v^0) + \sum_{j=1}^{n} \frac{\partial g_i}{\partial \xi_j} x_j + \sum_{j=1}^{l} \frac{\partial g_i}{\partial v_j} u_j \tag{1.10}$$

On substituting these expressions in (1.7) and (1.8) and noticing that

$$\dot{\xi}^0 = f(\xi^0, v^0); \qquad \xi^0(t_0) = c \tag{1.11}$$

$$\eta^0 = g(\xi^0, v^0) \tag{1.12}$$

the following equations are found for x and y:

$$\dot{x}_i = \sum_{j=1}^{n} \frac{\partial f_i}{\partial \xi_j} x_j + \sum_{j=1}^{l} \frac{\partial f_i}{\partial v_j} u_j; \qquad x_i(t_0) = 0 \tag{1.13}$$

$$y_i = \sum_{j=1}^{n} \frac{\partial g_i}{\partial \xi_j} x_j + \sum_{j=1}^{l} \frac{\partial g_i}{\partial v_j} u_j \tag{1.14}$$

The partial derivatives are evaluated for $\xi = \xi^0$, $v = v^0$, and so in general become functions of time. Equations (1.13) and (1.14) are therefore the same as (1.3) and (1.4). A similar result is obtained if f and g depend explicitly on t, as well as on ξ, v.

Transform methods lose most of their usefulness when we pass from equations (1.1), (1.2) to equations (1.3), (1.4). It is therefore necessary to deal directly with the differential equations. Because it is now more difficult to manipulate the equations, we assume that the differential equations are already in state-space form, as shown in (1.3), (1.4). For these reasons, and because (1.3), (1.4) permit more complicated behaviour than (1.1), (1.2), the theory of linear time-dependent systems is less complete than that of linear constant systems.

In this chapter we shall consider only differential systems. Time-dependent linear systems governed by difference equations can also be defined. They seem, however, to be less important in applications: at all events they have figured less extensively in the literature of the subject.

2 Transition matrices

If A, B, u are continuous functions of t on $[\alpha, \beta]$ it is known (Chapter 1, Theorem 7.2) that equation (1.3) has a unique solution passing through any given $x(t_0) = c$, where $t_0 \in [\alpha, \beta]$. This solution is written $\phi(t; c, t_0)$ or more briefly $\phi(t)$, is continuous on $[\alpha, \beta]$, and has continuous first derivative on (α, β). It follows that the matrix differential equation

$$\dot{\Psi}(t) = A(t)\Psi(t); \qquad \Psi(t_0) = I_n \tag{2.1}$$

has a unique solution Ψ on $[\alpha, \beta]$, because if Ψ is written in terms of its columns

$$\Psi = (\psi^1, \psi^2, \ldots, \psi^n) \tag{2.2}$$

equation (2.1) is equivalent to n equations of the form (1.3) with $u = 0$,

$$\dot{\psi}^k(t) = A(t)\psi^k(t); \qquad \psi_i^k(t_0) = \delta_{ik}, \quad i, k = 1, 2, \ldots, n \tag{2.3}$$

where δ_{ik} is the Kronecker delta (Chapter 1, Section 1). The matrix $\Psi(t)$ is called the *transition matrix* and it is known (Chapter 1, Section 7) to be non-singular for all $t \in [\alpha, \beta]$. If t_0 is understood, Ψ may be written $\Psi(t)$, as above, but more generally the dependence on t and t_0 is emphasized by writing $\Psi(t, t_0)$.

Theorem 2.1 If A is a continuous function of t on $[\alpha, \beta]$, and $t_0, t_1 \in [\alpha, \beta]$, then

$$\Psi(t_1, t_0) = \Psi^{-1}(t_0, t_1) \tag{2.4}$$

Proof Let H be a non-singular matrix, and define a matrix Φ by

$$\Phi(t) = \Psi(t, t_1)H \tag{2.5}$$

Then Φ is continuous on $[\alpha, \beta]$, and by (2.1) we have for $t \in (\alpha, \beta)$,

$$\dot{\Phi}(t) = A(t)\Phi(t); \qquad \Phi(t_0) = \Psi(t_0, t_1)H \tag{2.6}$$

Choose

$$H = \Psi^{-1}(t_0, t_1) \tag{2.7}$$

which exists because $\Psi(t_0, t_1)$ is non-singular. Then Φ satisfies

$$\dot{\Phi}(t) = A(t)\Phi(t); \qquad \Phi(t_0) = I_n \tag{2.8}$$

and so by definition is equal to $\Psi(t, t_0)$. Then on putting $t = t_1$, (2.5) gives (2.4).

Corollary $\Psi(t_1, t_0)$ is a continuous function of t_0 on $[\alpha, \beta]$.

Proof $\Psi(t_0, t_1)$ is a continuous function of t_0 on $[\alpha, \beta]$ by the properties of $\phi(t)$ stated before. Then the result follows from (2.4).

Theorem 2.2 If A is a continuous function of t on $[\alpha, \beta]$, and $t_0, t_1, t_2 \in [\alpha, \beta]$, then

$$\Psi(t_2, t_1) = \Psi(t_2, t_0)\Psi(t_0, t_1) \tag{2.9}$$

Proof The matrix

$$\Phi(t) = \Psi(t, t_0)\Psi(t_0, t_1) \tag{2.10}$$

satisfies, by (2.1) and Theorem 2.1,

$$\dot{\Phi}(t) = A(t)\Phi(t); \qquad \Phi(t_1) = I_n \tag{2.11}$$

whence by definition

$$\Phi(t) = \Psi(t, t_1) \tag{2.12}$$

On putting $t = t_2$, (2.10) then gives (2.9).

In previous chapters it was supposed, in accordance with the requirements of Laplace transformation, that variables were defined for all finite t. Physical variables were assumed in general to belong to D_R or to D_R^1. Similar assumptions will be made here, and as before no attempt will be made to give the most general results. The assumptions chosen will be those which are sufficient for most needs and which can be handled with a limited analytical background.

Theorem 2.3 Let A be continuous for all finite t, and let B, u belong to D_R. Then the solution of (1.3) which takes the value c at $t = t_0$ is

$$x(t) = \Psi(t, t_0)c + \int_{t_0}^{t} \Psi(t, \tau)B(\tau)u(\tau)\, d\tau \tag{2.13}$$

The solution so defined is continuous for all finite t and belongs to D_R^1.

Proof By assumption, B and u are in D_R, and so therefore is Bu. We may consequently write, for t in any bounded open interval J,

$$B(t)u(t) = \sum_{i=1}^{p} v_i U(t - p_i) + v(t) \tag{2.14}$$

where v is continuous on J. Let $[\alpha, \beta]$ be a closed interval in J containing t_0, and define $\Psi(t, t_0)$ for $t \in [\alpha, \beta]$ as above. Then if

$$x_0(t) = \Psi(t, t_0)c \tag{2.15}$$

we find from (2.1), for $t \in (\alpha, \beta)$,

$$\dot{x}_0(t) = A(t)\Psi(t, t_0)c = A(t)x_0(t) \tag{2.16}$$

Secondly consider

$$x_v(t) = \int_{t_0}^{t} \Psi(t, \tau)v(\tau)\, d\tau \tag{2.17}$$

$$= \Psi(t, t_0) \int_{t_0}^{t} \Psi(t_0, \tau)v(\tau)\, d\tau \tag{2.18}$$

which because v is continuous on $[\alpha, \beta]$ gives for $t \in (\alpha, \beta)$

$$\dot{x}_v(t) = A(t)\Psi(t, t_0)\int_{t_0}^{t} \Psi(t_0, \tau)v(\tau)\, d\tau + \Psi(t, t_0)\Psi(t_0, t)v(t) \tag{2.19}$$

$$= A(t)x_v(t) + v(t) \tag{2.20}$$

Thirdly consider

$$x_i(t) = \int_{t_0}^{t} \Psi(t, \tau)v_i U(\tau - p_i)\, d\tau \tag{2.21}$$

$$= \begin{cases} 0, & t \leqslant p_i \\ \int_{p_i}^{t} \Psi(t, \tau)v_i\, d\tau, & t > p_i \end{cases} \tag{2.22}$$

which is continuous for all $t \in [\alpha, \beta]$ and satisfies, for all $t \in (\alpha, \beta)$ except $t = p_i$,

$$\dot{x}_i(t) = A(t)x_i(t) + v_i U(t - p_i) \tag{2.23}$$

From this it follows that

$$x_i(t) - v_i(t - p_i)U(t - p_i) \tag{2.24}$$

has a continuous derivative on (α, β), whence $x_i \in D_R^1$ on (α, β).

Now putting

$$x(t) = x_0(t) + x_v(t) + \sum_{i=1}^{p} x_i(t) \tag{2.25}$$

we find that x is continuous on $[\alpha, \beta]$ and is in D_R^1 on (α, β). For all $t \in (\alpha, \beta)$ except $t = p_i$, $i = 1, 2, \ldots, p$,

$$\dot{x}(t) = A(t)x(t) + B(t)u(t) \tag{2.26}$$

That is to say, x defined by (2.13) is the solution of (1.3) on $[\alpha, \beta]$. As J, and $[\alpha, \beta]$, can be chosen arbitrarily large, this completes the proof.

Corollary If A is continuous for all finite t; B, C, D, u are in D_R; and $x(t_0) = 0$; then (1.3), (1.4) have the solution

$$y(t) = \int_{t_0}^{t} C(t)\Psi(t, \tau)B(\tau)u(\tau)\, d\tau + D(t)u(t) \tag{2.27}$$

and y so defined is in D_R. If C, D, u are in D_R^1, then y is in D_R^1.

Proof Equation (2.27) is

$$y(t) = C(t)x(t) + D(t)u(t) \tag{2.28}$$

where x is given by (2.13). As the product of two functions, each in D_R (resp. D_R^1), is also in D_R (resp. D_R^1), the result follows.

3 Step response and impulse response

The solution of equations (1.3) and (1.4) can be expressed in a useful alternative form. First define the *step response* H by

$$H(t, t_0) = \int_{t_0}^{t} C(t)\Psi(t, \tau)B(\tau)U(\tau - t_0)\, d\tau + D(t)U(t - t_0) \tag{3.1}$$

Then we have the following result.

Theorem 3.1 Let A be continuous for all finite t; $B \in D_R$; $C, D, u \in D_R^1$; $x(t_0) = 0$. Then the solution of (1.3), (1.4) is

$$y(t) = \int_{t_0}^{t} H(t, \tau)\, du(\tau) \tag{3.2}$$

Proof The system (1.3), (1.4) satisfies the following conditions:

(i) It is linear. For if the response to u_1 is y_1, and the response to u_2 is y_2, it follows from (2.27) that the response to $\alpha_1 u_1 + \alpha_2 u_2$ is $\alpha_1 y_1 + \alpha_2 y_2$.

(ii) It is non-anticipative. For if $u(t) = 0$, $t \leqslant t_0$, then (2.27) shows that $y(t) = 0$, $t \leqslant t_0$.

(iii) If $u(t) = 0$, $t \leqslant t_0$, and $\|u(t)\| < M_1$, $t_0 \leqslant t \leqslant T$, then $\|y(t)\| < M_2(M_1, T)$, $t_0 \leqslant t \leqslant T$. For

$$\|y(t)\| \leqslant \int_{t_0}^{t} \|C(t)\|\, \|\Psi(t, \tau)\|\, \|B(\tau)\|\, \|u(\tau)\|\, d\tau + \|D(t)\|\, \|u(t)\| \tag{3.3}$$

Then $\Psi(t, \tau)$ is continuous as a function of τ on the compact interval $[t_0, T]$, and so $\|\Psi(t, \tau)\|$ is bounded on this interval. The norms of B, C, D are also bounded on this interval because they are in D_R.

(iv) The matrix $H(t, t_0)$ is in D_L^1 as a function of t_0. For certainly $D(t)U(t - t_0)$ satisfies this condition, and we may write, for τ in an open interval $J \supset [t_0, T]$ and for fixed t,

$$\Psi(t, \tau)B(\tau) = \sum_{i=1}^{p} Q_i U(\tau - p_i) + Q(\tau) \tag{3.4}$$

where Q is continuous on J. Then if the integral in (3.1) is $I(t, t_0)$, and if $t > t_0$,

$$-\frac{\partial^+}{\partial t_0} I(t, t_0) = -\lim_{h\to 0+} \frac{1}{h} C(t)\left\{\int_{t_0+h}^{t} \left[\sum_{i=1}^{p} Q_i U(\tau - p_i) + Q(\tau)\right] d\tau - \int_{t_0}^{t} \left[\sum_{i=1}^{p} Q_i U(\tau - p_i) + Q(\tau)\right] d\tau\right\}$$

$$= \lim_{h\to 0+} \frac{1}{h} C(t) \int_{t_0}^{t_0+h} \left[\sum_{i=1}^{p} Q_i U(\tau - p_i) + Q(\tau)\right] d\tau \quad (3.5)$$

On noting that Q is continuous and that

$$\lim_{h\to 0+} \frac{1}{h} \int_0^h U(\tau)\, d\tau \quad (3.6)$$

is 1, and not 0, we find that (3.5) gives

$$-\frac{\partial^+}{\partial t_0} I(t, t_0) = C(t)\left\{\sum_{i=1}^{p} Q_i[1 - U(p_i - t_0] + Q(t_0)\right\} \quad (3.7)$$

When $t \leqslant t_0$, a similar calculation gives $-\partial^+ I(t, t_0)/\partial t_0 = 0$, whence for all t, t_0 in J,

$$-\frac{\partial^+}{\partial t_0} I(t, t_0) = C(t)\left\{\sum_{i=1}^{p} Q_i[1 - U(p_i - t_0)] + Q(t_0)\right\}U(t - t_0) \quad (3.8)$$

A similar calculation shows that $-\partial^- I(t, t_0)/\partial t_0$ also exists for all t, t_0 in J, and differs from $-\partial^+ I(t, t_0)/\partial t_0$ only at the points p_i. Since I is a continuous function of t_0, it readily follows that $I \in D_L^1$ as a function of t_0.

Properties (i) to (iv) imply (Chapter 1, Theorem 10.1) that the output y is given by (3.2).

If $D = 0$ we define the *impulse response* h by

$$h(t, t_0) = -\frac{\partial^+}{\partial t_0} H(t, t_0) = -\frac{\partial^+}{\partial t_0} I(t, t_0) \quad (3.9)$$

which is given explicitly by (3.8), is in D_L as a function of t_0, and differs from $C(t)\Psi(t, t_0)B(t_0)U(t - t_0)$ only at the discontinuities p_i of B. If $D(t)$ is not identically zero, we define $h(t, t_0)$ in the distribution sense (R & S, p. 221) by

$$h(t, t_0) = -\frac{\partial^+}{\partial t_0} \int_{t_0}^{t} C(t)\Psi(t, \tau)B(\tau)\, U(\tau - t_0)\, d\tau + D(t)\delta(t - t_0) \quad (3.10)$$

where $\delta(t - t_0)$ is a delta function at $t = t_0$. With this interpretation of h, equation (3.2) becomes

$$y(t) = \int_{t_0}^{t} h(t, \tau)u(\tau)\, d\tau \quad (3.11)$$

It follows from (3.9) that if $D = 0$ and B, C are continuous, then $h(t, t_0)$ is continuous as a function of t and of t_0 (separately) for all t_0 and $t > t_0$. If a matrix h is given with this property, is it the impulse response of some system (1.3), (1.4) with $D = 0$? The answer to this question is provided by the following theorem.

Theorem 3.2 Let $h(t, t_0)$ be an $m \times l$ matrix which is a continuous function of t and t_0 (separately) for all t_0 and $t > t_0$ and is zero for $t \leqslant t_0$. Then $h(t, t_0)$ is the impulse response of some system (1.3), (1.4) with $D = 0$ and A, B, C continuous for all finite t, and with finite order n, if and only if

$$h(t, t_0) = P(t)Q(t_0)U(t - t_0) \tag{3.12}$$

for all t, t_0. Here P, Q are continuous for all finite t, t_0 and have dimension $m \times n$, $n \times l$ respectively.

Proof If h arises from (1.3), (1.4) with $D = 0$ and B, C continuous, (3.9) shows that

$$h(t, t_0) = C(t)\Psi(t, t_0)B(t_0)U(t - t_0) \tag{3.13}$$

$$= C(t)\Psi(t, t_1)\Psi(t_1, t_0)B(t_0)U(t - t_0) \tag{3.14}$$

$$= P(t)Q(t_0)U(t - t_0) \tag{3.15}$$

where t_1 is any fixed time. Conversely, if (3.12) is true, put

$$\dot{x}(t) = Q(t)u(t) \tag{3.16}$$

$$y(t) = P(t)x(t) \tag{3.17}$$

This system has transition matrix

$$\Psi(t, t_0) = I_n \tag{3.18}$$

and as in (3.13)

$$h(t, t_0) = P(t)Q(t_0)U(t - t_0) \tag{3.19}$$

4 Operational system matrices

Because transform methods are generally unprofitable for time-dependent systems, the system matrix cannot be defined as in Chapter 2. However, we may define an operational form of the system matrix by writing equations (1.3), (1.4) in the form

$$\begin{pmatrix} pI_n - A(t) & B(t) \\ -C(t) & D(t) \end{pmatrix} \begin{pmatrix} x(t) \\ -u(t) \end{pmatrix} = \begin{pmatrix} 0 \\ -y(t) \end{pmatrix} \tag{4.1}$$

Here p represents the differential operator: D is not used because it could be confused with the matrix $D(t)$. The system matrix

$$P(p, t) = \begin{pmatrix} pI_n - A(t) & B(t) \\ -C(t) & D(t) \end{pmatrix} \tag{4.2}$$

will be said to be in state-space form, and also to be in operational form.

Theorem 4.1 Let $Q(t)$ be an $n \times n$ matrix which is non-singular and has continuous derivative $\dot{Q}(t)$ for all finite t. Then the system matrix

$$P_1(p, t) = \begin{pmatrix} Q^{-1}(t) & 0 \\ 0 & I_m \end{pmatrix}\begin{pmatrix} pI_n - A(t) & B(t) \\ -C(t) & D(t) \end{pmatrix}\begin{pmatrix} Q(t) & 0 \\ 0 & I_l \end{pmatrix} \tag{4.3}$$

gives rise to the same step response matrix $H(t, t_0)$ as $P(p, t)$ given in (4.2).

Proof Multiply both sides of (4.1) by $\begin{pmatrix} Q^{-1} & 0 \\ 0 & I_m \end{pmatrix}$. The result can be written

$$P_1(p, t)\begin{pmatrix} Q^{-1}x \\ -u \end{pmatrix} = \begin{pmatrix} 0 \\ -y \end{pmatrix} \tag{4.4}$$

which shows that the systems corresponding to P and to P_1 each give the same output y for a given input u and corresponding initial conditions. On choosing the matrix input $I_l U(t - t_0)$ and zero initial conditions at $t = t_0$ the theorem follows.

A direct evaluation of (4.4) with $Q^{-1}x$ replaced by x_1 shows that P_1 defined by (4.3) is to be interpreted as

$$P_1(p, t) = \begin{pmatrix} pI_n - Q^{-1}AQ + Q^{-1}\dot{Q} & Q^{-1}B \\ -CQ & D \end{pmatrix} \tag{4.5}$$

The condition in Theorem (4.1) that $\dot{Q}$ should be continuous for all finite t was imposed in order to ensure that $Q^{-1}AQ - Q^{-1}\dot{Q}$ has the same continuity properties as A.

EXERCISE 4.1 Show that the time-dependent system (with $n = 1$)

$$\dot{x}_1(t) = -\frac{\cos t}{2 + \sin t}x_1(t) + \frac{1}{2 + \sin t}u(t) \tag{4.6}$$

$$y_1(t) = (2 + \sin t)x_1(t) \tag{4.7}$$

has the same impulse response $h(t, t_0) = h(t - t_0)$ as the constant system

$$\dot{x}(t) = u(t) \tag{4.8}$$

$$y(t) = x(t) \tag{4.9}$$

The translation of system matrices into operational form can be pursued further, as in the following examples.

Example 4.1 Let $T(p, t)$ be an $r \times r$ matrix having as its elements polynomials in the operator p, with coefficients which are suitably smooth functions of t. Then T can be written

$$T(p, s) = T_0(t) + T_1(t)p + \cdots + T_q(t)p^q \tag{4.10}$$

Let $U(p, t)$, $V(p, t)$, $W(p, t)$ be matrices, respectively $r \times l$, $m \times r$, $m \times l$, which are defined in a similar way. Then if u is a sufficiently smooth function of t,

$$\begin{pmatrix} T(p, s) & U(p, s) \\ -V(p, s) & W(p, s) \end{pmatrix} \begin{pmatrix} \xi(t) \\ -u(t) \end{pmatrix} = \begin{pmatrix} 0 \\ -y(t) \end{pmatrix} \tag{4.11}$$

defines a set of differential equations which may represent a physical system.

Example 4.2 Let $M(p, t)$, $N(p, t)$, $X(p, t)$, $Y(p, t)$ be matrices, respectively $r \times r$, $r \times r$, $m \times r$, $r \times l$, having as their elements polynomials in p with the coefficients suitably smooth functions of time. Let M and N have their determinants independent of p and non-zero for all finite t. Then the system corresponding to

$$P_1(p, t) = \begin{pmatrix} M & 0 \\ X & I_m \end{pmatrix} \begin{pmatrix} T & U \\ -V & W \end{pmatrix} \begin{pmatrix} N & Y \\ 0 & I_l \end{pmatrix} \tag{4.12}$$

gives rise to the same step response as the system corresponding to

$$P(p, t) = \begin{pmatrix} T & U \\ -V & W \end{pmatrix} \tag{4.13}$$

For on multiplying both sides of (4.11) by

$$\begin{pmatrix} M & 0 \\ X & I_m \end{pmatrix} \tag{4.14}$$

and writing

$$\begin{pmatrix} M^{-1}\xi - M^{-1}Yu \\ -u \end{pmatrix} = \begin{pmatrix} \xi_1 \\ -u \end{pmatrix} \tag{4.15}$$

we obtain

$$P_1(p, t)\begin{pmatrix} \xi_1(t) \\ -u(t) \end{pmatrix} = \begin{pmatrix} 0 \\ -y(t) \end{pmatrix} \tag{4.16}$$

This shows that if (4.11) gives an output y for given u, then (4.16) gives the same output y when the initial conditions correspond and the input is the given u. In particular we may choose an input $I_l U(t - t_0)$, with zero initial conditions at $t = t_0$, taking derivatives in the distribution sense.

EXERCISE 4.2 Show how to interpret a product such as $M(p, t)U(p, t)$ occurring in (4.12). [If

$$M(p, t) = M_0(t) + M_1(t)p + \cdots + M_q(t)p^q \tag{4.17}$$

$$U(p, t) = U_0(t) + U_1(t)p + \cdots + U_s(t)p^s \tag{4.18}$$

then in the evaluation of (4.16) there will occur

$$\big(M_0(t) + M_1(t)p + \cdots + M_q(t)p^q\big)\big(U_0(t) + U_1(t)p + \cdots + U_s(t)p^s\big)u(t) \tag{4.19}$$

Then for example

$$(M_i p^i)(U_j p^j)u = M_i\left[(p^i U_j)(p^j u) + \binom{i}{1}(p^{i-1}U_j)(p^{j+1}u) + \binom{i}{2}(p^{i-2}U_j)(p^{j+2}u) + \cdots + U_j(p^{j+i}u)\right] \tag{4.20}$$

where

$$\binom{i}{k} = \frac{i!}{k!(i-k)!}$$

In this way all the products occurring in (4.19) can be evaluated.]

EXERCISE 4.3 Examine how a triple product such as $M(p, t)T(p, t)N(p, t)$ in (4.12) is to be interpreted.

EXERCISE 4.4 On the interval [1, 2], the equations

$$\left.\begin{aligned} t\dot{z}_1 &= -z_2 - tu; & z_1(1) &= 0 \\ \dot{z}_2 &= -z_1 - t\dot{u}; & z_2(1) &= 0 \\ y &= tz_1 + z_2 \end{aligned}\right\} \tag{4.21}$$

define y for any u with continuous first derivative. The corresponding operational system matrix is

$$P(p, t) = \left(\begin{array}{cc|c} tp & 1 & -t \\ 1 & p & -tp \\ \hline -t & -1 & 0 \end{array}\right) \tag{4.22}$$

Write out the differential equations corresponding to

$$P_1(p, t) = \begin{pmatrix} 1 & -tp & 0 \\ 0 & t & 0 \\ 1 & -tp & 1 \end{pmatrix} P(p, t) \begin{pmatrix} 1 & 0 & 0 \\ 0 & 1 & t \\ 0 & 0 & 1 \end{pmatrix} \tag{4.23}$$

and show that they lead to the same step response as (4.21). [The equations are, after some simplification,

$$\left.\begin{aligned} \dot{\zeta}_2 &= -\zeta_1 + u \\ t\ddot{\zeta}_2 &= \zeta_2 + t\dot{u} \\ y &= t\zeta_1 + \zeta_2 - tu \end{aligned}\right\} \tag{4.24}$$

By putting $z_1 = \zeta_1$, $z_2 = \zeta_2 - tu$, these can be derived from (4.21).]

5 Controllability

In the treatment of linear constant differential systems, a fundamental role was played by common divisors and the idea of relative primeness. Controllability (p.s.) and observability were then shown as two among many consequences of the latter property. For time-dependent systems the algebraic properties are lost, but the ideas of controllability (p.s.) and observability still apply. These ideas have played an important part in motivating and guiding the historical development of the subject. They are therefore introduced here at an earlier stage than is logically necessary. [Compare the development in Chapters 2 to 5.]

The system (1.3) will be called *controllable (p.s.t.o.) on* $[t_0, t_1]$ if, given an arbitrary state c_0 at t_0, there exists an input $u \in D_R$ which transfers the state to the origin at t_1. *Controllability (p.s.f.o.) on* $[t_0, t_1]$ means that given an arbitrary state c_1 there exists an input $u \in D_R$ which transfers the state from the origin at t_0, to c_1 at t_1. The system will be called *controllable (p.s.) on* $[t_0, t_1]$ if, given two arbitrary states c_0, c_1, there exists an input $u \in D_R$ which transfers the state from c_0 at t_0, to c_1 at t_1.

Theorem 5.1 Let A be continuous for all finite t; $B \in D_R$; and let the system (1.3) be controllable (p.s.) [resp. (p.s.t.o.), (p.s.f.o.)] on $[t_0, t_1]$. Then the system is controllable (p.s.) [resp. (p.s.t.o.), (p.s.f.o.)] on any interval $[t_0, t_2]$ with $t_2 > t_1$.

Proof Because A is continuous there is (Chapter 1, Theorem 7.2) a unique solution of (1.3) with $u = 0$ which passes through any given c_2 at t_2 and is defined for all finite t. Call this free solution $\phi_0(t; c_2, t_2)$.

(a) If $c_2 = 0$, then $\phi_0(t_1; 0, t_2) = 0$. Let the system be controllable (p.s.t.o.). Then the input u on $[t_0, t_1]$ which carries c_0 at t_0 to 0 at t_1, if followed by $u = 0$ on $(t_1, t_2]$, will carry c_0 at t_0 to 0 at t_2.

(b) Let the system be controllable (p.s.) [resp. controllable (p.s.f.o.)]. Let an arbitrary state c_2 at t_2 be specified. Then there exists an input u on $[t_0, t_1]$ which drives c_0 at t_0 [resp. 0 at t_0] to $\phi_0(t_1; c_2, t_2)$ at t_1. This input on $[t_0, t_1]$ followed by $u = 0$ on $(t_1, t_2]$ drives c_0 at t_0 [resp. 0 at t_0] to c_2 at t_2.

Theorem 5.2 Controllability (p.s.) on $[t_0, t_1]$ implies controllability (p.s.t.o.) on $[t_0, t_1]$ and controllability (p.s.f.o.) on $[t_0, t_1]$. Controllability (p.s.t.o.) on $[t_0, t_1]$ and controllability (p.s.f.o.) on $[t_1, t_2]$ together imply controllability (p.s.) on $[t_0, t_2]$ when $t_0 < t_1 < t_2$.

Proof The first part follows from the definition of controllability (p.s.). The second part is obvious.

Theorem 5.3 Let A be continuous for all finite t and let B be in D_R. The system (1.3) is controllable (p.s.) on $[t_0, t_1]$ if and only if the *controllability matrix* $W(t_0, t_1)$ defined by

$$W(t_0, t_1) = \int_{t_0}^{t_1} \Psi(t_0, \tau)B(\tau)B^{\mathrm{T}}(\tau)\Psi^{\mathrm{T}}(t_0, \tau)\, d\tau \tag{5.1}$$

is non-singular. The input which then transfers any state c_0 at t_0 to c_1 at t_1 can be chosen in D_R.

Proof If $W(t_0, t_1)$ is non-singular, let c be the solution of

$$W(t_0, t_1)c = \Psi(t_0, t_1)c_1 - c_0 \tag{5.2}$$

and consider the input

$$u(t) = B^{\mathrm{T}}(t)\Psi^{\mathrm{T}}(t_0, t)c \tag{5.3}$$

which is in D_R. The corresponding solution passing through c_0 at t_0 is

$$\phi_u(t; c_0, t_0) = \Psi(t, t_0)c_0 + \int_{t_0}^{t} \Psi(t, \tau)B(\tau)u(\tau)\, d\tau \tag{5.4}$$

whence, using Theorem 2.2,

$$\phi_u(t_1; c_0, t_0) = \Psi(t_1, t_0)c_0 + \int_{t_0}^{t_1} \Psi(t_1, t_0)\Psi(t_0, \tau)B(\tau)B^{\mathrm{T}}(\tau)\Psi^{\mathrm{T}}(t_0, \tau)c\, d\tau \tag{5.5}$$

$$= \Psi(t_1, t_0)c_0 + \Psi(t_1, t_0)W(t_0, t_1)c \tag{5.6}$$

$$= c_1 \tag{5.7}$$

which proves sufficiency.

If $W(t_0, t_1)$ is singular, choose $c_0 \neq 0$ so that $W(t_0, t_1)c_0 = 0$. Suppose, contrary to what is to be proved, that there exists a $u \in D_R$ which transfers the state from c_0 at t_0 to the origin at t_1. Then

$$0 = \Psi(t_1, t_0)c_0 + \int_{t_0}^{t_1} \Psi(t_1, \tau)B(\tau)u(\tau)\, d\tau \tag{5.8}$$

$$c_0 = -\Psi(t_0, t_1)\int_{t_0}^{t_1} \Psi(t_1, \tau)B(\tau)u(\tau)\, d\tau \tag{5.9}$$

and so

$$0 < c_0^{\mathrm{T}}c_0 = -\int_{t_0}^{t_1} c_0^{\mathrm{T}}\Psi(t_0, \tau)B(\tau)u(\tau)\, d\tau \tag{5.10}$$

But from the way in which c_0 was chosen

$$0 = c_0^{\mathrm{T}}W(t_0, t_1)c_0 = \int_{t_0}^{t_1} c_0^{\mathrm{T}}\Psi(t_0, \tau)B(\tau)B^{\mathrm{T}}(\tau)\Psi^{\mathrm{T}}(t_0, \tau)c_0\, d\tau \tag{5.11}$$

$$= \int_{t_0} \|c_0^{\mathrm{T}}\Psi(t_0, \tau)B(\tau)\|^2\, d\tau \tag{5.12}$$

The integrand in (5.12) is continuous on the left and non-negative. Let it take the value $2\varepsilon > 0$ at some $t_2 \in (t_0, t_1]$. Then there exists $\delta > 0$ such that the integrand exceeds ε for $t \in [t_2 - \delta, t_2] \cap [t_0, t_1]$, which is an interval not less than $\delta_1 = \min(\delta, t_2 - t_0)$ in length. This implies that the integral (5.12) is not less than $\varepsilon\delta_1 > 0$, contradicting (5.11). Hence the integrand in (5.12) is zero on $(t_0, t_1]$. The integrand in (5.10) therefore differs from zero, if at all, only at the single point $\tau = t_0$ and the integrand may be put equal to zero at this single point without changing the value of the integral (Chapter 1, Section 6) which is therefore zero. But this shows that (5.10) is a contradiction, and so there exists no $u \in D_R$ which transfers the state from c_0 at t_0 to 0 at t_1. This proves necessity.

Theorem 5.4 Let A be continuous for all finite t and let B be in D_R. A state c_0 at t_0 can be transferred to c_1 at t_1 if and only if $\Psi(t_0, t_1)c_1 - c_0$ lies in the *controllable subspace* Ω spanned by the columns of $W(t_0, t_1)$. Such transfer can be achieved by u which is in D_R.

Proof By the conditions of the theorem we may choose c such that

$$W(t_0, t_1)c = \Psi(t_0, t_1)c_1 - c_0 \tag{5.13}$$

Then it follows as in Theorem 5.3 that the input defined by (5.3) transfers the state from c_0 at t_0 to c_1 at t_1. The proof of necessity is the same as in Theorem 5.3.

EXERCISE 5.1 Show that $W(t_0, t_1)$ is symmetric.

EXERCISE 5.2 Show that $W(t_0, t_1)$ is positive semidefinite for $t_1 \geqslant t_0$.

EXERCISE 5.3 Show that $W(t_0, t_1)$ can be written $H^T H$ if $t_1 \geqslant t_0$. [A symmetric real matrix can be diagonalized by orthogonal similarity (Chapter 1, Section 1.1) and has non-negative eigenvalues if it is positive semidefinite.]

EXERCISE 5.4 Show that rank $W(t_0, t_1)$ is a non-decreasing function of t_1 for $t_1 > t_0$. [Consider $v^T W(t_0, t_1) v$ as a function of t_1.]

The link between the above results and those proved in Chapter 5, Section 2.3 is provided by the following.

Theorem 5.5 If A, B are constant, the subspace Ω spanned by the columns of $W(t_0, t_1)$ is the same, for any t_0 and $t_1 > t_0$, as the subspace spanned by the columns of $(B\ AB\ \ldots\ A^{q-1}B)$. Here q may be any integer not less than the degree of the minimal polynomial of A.

Proof When A and B are constant

$$\Psi(t_0, \tau) = e^{A(t_0 - \tau)} \tag{5.14}$$

and

$$W(t_0, t_1) = \int_{t_0}^{t_1} e^{A(t_0 - \tau)} B B^T e^{A^T(t_0 - \tau)}\, d\tau \tag{5.15}$$

Let the n-vector v lie in the null space N_1 of $(B\ AB\ \ldots\ A^{q-1}B)^T$; that is

$$\begin{pmatrix} B^T \\ B^T A^T \\ \vdots \\ B^T (A^T)^{q-1} \end{pmatrix} v = 0 \tag{5.16}$$

Then transposing (5.16) we find that

$$v^T (B \quad AB \quad \ldots \quad A^{q-1}B) = 0 \tag{5.17}$$

so that v is orthogonal to every column of $(B\ AB\ \ldots\ A^{q-1}B)$. Conversely, every vector v orthogonal to all of these columns is in the null space N_1. That is, the subspace spanned by the columns of $(B\ AB\ \ldots\ A^{q-1}B)$ is the orthogonal complement of N_1.

In the same way the subspace spanned by the columns of $W(t_0, t_1)$ is the orthogonal complement of the null space N_2 of $W^T(t_0, t_1)$. Because $W(t_0, t_1)$ is symmetric, from (5.15), N_2 is also the null space of $W(t_0, t_1)$. If we can show that the null spaces N_1 and N_2 coincide, the theorem will be proved.

Let w lie in N_2. Then (5.15) gives

$$0 = w^{\mathrm{T}} W(t_0, t_1) w = \int_{t_0}^{t_1} w^{\mathrm{T}} e^{A(t_0 - \tau)} B B^{\mathrm{T}} e^{A^{\mathrm{T}}(t_0 - \tau)} w \, d\tau \tag{5.18}$$

$$= \int_{t_0}^{t_1} \| B^{\mathrm{T}} e^{A^{\mathrm{T}}(t_0 - \tau)} w \|^2 \, d\tau \tag{5.19}$$

Because the integrand is a non-negative continuous function of τ, it follows that

$$B^{\mathrm{T}} e^{A^{\mathrm{T}}(t_0 - \tau)} w = 0 \tag{5.20}$$

for all $\tau \in [t_0, t_1]$. Differentiate (5.20) $0, 1, \ldots, q - 1$ times with respect to τ and then put $\tau = t_0$, to obtain

$$\left.\begin{aligned} B^{\mathrm{T}} w &= 0 \\ B^{\mathrm{T}} A^{\mathrm{T}} w &= 0 \\ &\vdots \\ B^{\mathrm{T}} (A^{\mathrm{T}})^{q-1} w &= 0 \end{aligned}\right\} \tag{5.21}$$

which shows that $w \in N_1$.

Now let v lie in N_1. If the minimal polynomial of A is

$$\alpha_0 I + \alpha_1 A + \cdots + A^k = 0 \tag{5.22}$$

with $k \leqslant q$, then every one of $A^k, A^{k+1}, \ldots$ can be expressed as a linear combination of $I, A, \ldots, A^{k-1}$. Since all of $B^{\mathrm{T}} v, B^{\mathrm{T}} A^{\mathrm{T}} v, \ldots, B^{\mathrm{T}} (A^{\mathrm{T}})^{k-1} v$ are zero, it follows that $B^{\mathrm{T}} (A^{\mathrm{T}})^k v, B^{\mathrm{T}} (A^{\mathrm{T}})^{k+1} v, \ldots$ are also zero. From the Taylor series expansion of $e^{A^{\mathrm{T}}(t_0 - \tau)}$ we then find that

$$B^{\mathrm{T}} e^{A^{\mathrm{T}}(t_0 - \tau)} v = 0 \tag{5.23}$$

for all τ, and so, as in (5.19),

$$v^{\mathrm{T}} W(t_0, t_1) v = 0 \tag{5.24}$$

Now W is symmetric and positive semidefinite, and can be written [Exercise 5.3]

$$W(t_0, t_1) = H^{\mathrm{T}} H \tag{5.25}$$

Equation (5.22) then gives

$$\| H v \|^2 = 0 \tag{5.26}$$

whence $Hv = 0$ and so $W(t_0, t_1) v = 0$. That is, $v \in N_2$. Consequently N_1 and N_2 coincide, which completes the proof.

Corollary In Chapter 5, Theorem 2.4, u can be chosen continuous.

Proof This follows from Theorems 5.4 and 5.5, on putting B constant in equation (5.3).

EXERCISE 5.5 Let A, B be constant and define W_0 by

$$W_0 = (B \quad AB \quad \dots \quad A^{q-1}B)\begin{pmatrix} B^T \\ B^T A^T \\ \vdots \\ B^T(A^T)^{q-1} \end{pmatrix} \tag{5.27}$$

where q is as defined in Theorem 5.5. Show that for any t_0 and $t_1 > t_0$ the columns of W_0 and of $W(t_0, t_1)$ span the same subspace, and that W_0 and $W(t_0, t_1)$ have the same null space.

The heuristic importance of controllability (p.s.) is that if, for all $t_1 > t_0$, Ω in Theorem 5.4 has dimension less than the order of the system, then only states in Ω can be reached from $c_0 = 0$ at t_0. So far as the input–output behaviour is concerned, with this initial condition, it follows that a state space of smaller dimension should suffice. Similar remarks apply to observability. These ideas therefore suggest conditions which will ensure that the state space has the smallest possible dimension consistent with given input–output behaviour. Considerable difficulties arise, however, in making this intuitive notion precise, as will be seen in Section 7.

6 Observability

The idea of observability generalizes from time-invariant to time-dependent systems in much the same way as that of controllability (p.s.). Knowledge of u and of the matices A, B, C, D is assumed, so that if we are interested in deducing the state from measurements of the output y, the effect of u can always be eliminated. Accordingly we may as well deal with

$$\dot{x}(t) = A(t)x(t) \tag{6.1}$$

$$y(t) = C(t)x(t) \tag{6.2}$$

This system [or the full system (1.3), (1.4)] will be called *observable on* $[t_0, t_1]$ if, given y on this interval, we may deduce $x(t_0)$.

Theorem 6.1 The system (6.1), (6.2), with A, C continuous for all finite t is observable on $[t_0, t_1]$ if and only if the *observability matrix*

$$M(t_0, t_1) = \int_{t_0}^{t_1} \Psi^T(\tau, t_0)C^T(\tau)C(\tau)\Psi(\tau, t_0)\, d\tau \tag{6.3}$$

is non-singular.

Proof Equations (6.1), (6.2), give

$$y(t) = C(t)\Psi(t, t_0)x(t_0) \tag{6.4}$$

whence

$$\Psi^{\mathrm{T}}(t, t_0)C^{\mathrm{T}}(t)y(t) = \Psi^{\mathrm{T}}(t, t_0)C^{\mathrm{T}}(t)C(t)\Psi(t, t_0)x(t_0) \tag{6.5}$$

and

$$\int_{t_0}^{t_1} \Psi^{\mathrm{T}}(\tau, t_0)C^{\mathrm{T}}(\tau)y(\tau)\, d\tau = M(t_0, t_1)x(t_0) \tag{6.6}$$

so that if $M(t_0, t_1)$ is non-singular, $x(t_0)$ is determined by

$$x(t_0) = M^{-1}(t_0, t_1) \int_{t_0}^{t_1} \Psi^{\mathrm{T}}(\tau, t_0)C^{\mathrm{T}}(\tau)y(\tau)\, d\tau \tag{6.7}$$

If $M(t_0, t_1)$ is singular, choose any non-zero $x(t_0)$ such that

$$M(t_0, t_1)x(t_0) = 0 \tag{6.8}$$

Then if $y(t)$ is the corresponding output

$$\int_{t_0}^{t} \|y(\tau)\|^2\, d\tau = \int_{t_0}^{t_1} y^{\mathrm{T}}(\tau)y(\tau)\, d\tau \tag{6.9}$$

$$= \int_{t_0}^{t_1} x^{\mathrm{T}}(t_0)\Psi^{\mathrm{T}}(\tau, t_0)C^{\mathrm{T}}(\tau)C(\tau)\Psi(\tau, t_0)x(t_0)\, d\tau \tag{6.10}$$

$$= x^{\mathrm{T}}(t_0)M(t_0, t_1)x(t_0) \tag{6.11}$$

$$= 0 \tag{6.12}$$

Then because y is continuous, (6.12) implies that $y(t) = 0$ on $[t_0, t_1]$. Equation (6.4) shows that $x(t_0) = 0$ also gives $y(t) = 0$ on $[t_0, t_1]$, so that no $x(t_0)$ satisfying (6.8) can be distinguished from zero by knowledge of y on $[t_0, t_1]$.

Notice that if C had been assumed to belong to D_R or D_R^1, y could have had discontinuities on the right. Then from (6.12) we should have been able to conclude, as in the proof of Theorem 5.3, only that $y(t) = 0$ on $(t_0, t_1]$. The value of $y(t_0)$ might then have provided information about $x(t_0)$.

Example 6.1 Let $n = 1$, $\dot{x} = 0$, $C(t) = 1 - U(t)$. Then if $x(0) = c$, we have $x(t) = c$, $t \geqslant 0$, and $y(t) = [1 - U(t)]c$, so that by observing $y(0) = c$ we can identify $x(0)$. This is despite the fact that

$$M(0, t_1) = \int_0^{t_1} [1 - U(\tau)]^2\, d\tau \tag{6.13}$$

$$= 0 \tag{6.14}$$

for all $t_1 > 0$.

Example 6.1 discloses an asymmetry between the conditions we must place on B and C when investigating controllability (p.s.) and observability respectively. The asymmetry arises because Bu appears under the integral sign in (2.13), and the value of an integrand may be changed at any finite set of points without changing the value of the integral. On the other hand y is given directly by Cx. As we shall in fact observe y by noting its effect on another dynamical system, the asymmetry has no practical significance.

EXERCISE 6.1 Show that $M(t_0, t_1)$ satisfies the conditions given for $W(t_0, t_1)$ in Exercises 5.1 to 5.4.

EXERCISE 6.2 Show that if A, C are constant, then the subspace spanned by the columns of $M(t_0, t_1)$ for any t_0 and $t_1 > t_0$ is the same as the subspace spanned by the columns of $[C^T, A^TC^T, \ldots, (A^T)^{q-1}C^T]$, where q is any integer not less than the degree of the minimal polynomial of A.

7 Least order

For simplicity we will now assume that A, B, C in (1.3), (1.4) are continuous for all finite t, and $D = 0$, so that the system becomes

$$\dot{x}(t) = A(t)x(t) + B(t)u(t) \tag{7.1}$$

$$y(t) = C(t)x(t) \tag{7.2}$$

Then the impulse response matrix $h(t, t_0)$ can be written

$$h(t, t_0) = P(t)Q(t_0)U(t - t_0) \tag{7.3}$$

where P, Q are continuous for all finite t, t_0, and are respectively $m \times n$, $n \times l$ [Theorem 3.2]. The matrix $h(t, t_0)$ characterizes the input–output behaviour of the system, because (3.11) gives, for $x(t_0) = 0$,

$$y(t) = \int_{t_0}^{t} h(t, \tau)u(\tau)\, d\tau \tag{7.4}$$

Given $h(t, t_0)$ satisfying (7.3), we may ask whether there is an alternative decomposition

$$h(t, t_0) = P_0(t)Q_0(t_0)U(t - t_0) \tag{7.5}$$

with P_0, Q_0 respectively $m \times n_0$, $n_0 \times l$, and with $n_0 < n$. If so there exists a system (3.16), (3.17) with order $n_0 < n$ giving rise to $h(t, t_0)$, and therefore giving the same input–output behaviour as the original system (7.1), (7.2). On the other hand, if no decomposition (7.5) exists with $n_0 < n$, we shall say that the original system had *least order*.

These considerations clearly paraphrase questions which were considered for time-invariant systems in Chapter 3, Section 3. When A, B, C are constant, $h(t, t_0)$ becomes a function only of $t - t_0$, and the Laplace transform of $h(t - t_0)$ with $t_0 = 0$ is the transfer function matrix $G(s)$ (Chapter 1, Section 10).

Unfortunately, time-dependent systems offer great difficulties and no complete answer is known to the problem as stated. The difficulties arise from the fact that causality has to be imposed somewhat arbitrarily on the system (7.1), (7.2) by the introduction of $U(t - t_0)$ in (7.3). The easiest way to proceed is therefore to ask a weaker form of the above question in which causality does not intervene. To this end we first define the *weighting pattern* as the matrix PQ occurring in (7.3): it is equal to h for $t > t_0$, and is given explicitly in equation (3.13). The usual symbol for the weighting pattern is $W(t, t_0)$, but we shall avoid this in view of the conflict with equation (5.1). Then given P and Q in (7.3), we shall ask whether there exist P_0 and Q_0 as in (7.5) with $n_0 < n$, and such that $P_0(t)Q_0(t_0) = P(t)Q(t_0)$ for all t, t_0.

Our investigation will follow similar principles to those of the earlier investigation, though now the analysis must be conducted by other methods. First we shall give a procedure by which the order of a system can sometimes be reduced (compare Chapter 2, Section 4). When two systems cannot be further reduced in this way, and they give the same weighting pattern, we show that they have the same order. We also show that they are related by a transformation of the type given in equation (4.3).

In equation (7.3) the columns of $P(t)$ will be called linearly independent if there exists no non-zero constant n-vector v such that $P(t)v = 0$ for all t. If there exists such a vector, the columns of $P(t)$ will be called linearly dependent. The rows of $Q(t)$ will be called linearly independent if the columns of $Q^T(t)$ are linearly independent. With these definitions we have the following result.

Theorem 7.1 If either the columns of P or the rows of Q are linearly dependent, there exists a system of order $n_0 < n$ giving rise to the same weighting pattern PQ.

Proof Let there exist $v \neq 0$ such that $P(t)v = 0$ for all t. Construct a non-singular constant $n \times n$ matrix H having v as its last column. Then

$$P(t)Q(t_0) = P(t)HH^{-1}Q(t_0) \tag{7.6}$$

The last column of $P(t)H$ is zero for all t, and if we write

$$P(t)H = (P_1(t), 0) \tag{7.7}$$

$$H^{-1}Q(t_0) = \begin{pmatrix} Q_1(t_0) \\ q(t_0) \end{pmatrix} \tag{7.8}$$

where P_1, Q_1 are respectively $m \times (n - 1)$, $(n - 1) \times l$, it follows that

$$P(t)Q(t_0) = P_1(t)Q_1(t_0) \tag{7.9}$$

Then there exists a system

$$\dot{x}_1(t) = Q_1(t)u(t) \tag{7.10}$$

$$y(t) = P_1(t)x_1(t) \tag{7.11}$$

having weighting pattern PQ, and having order $n - 1$. This process can be repeated if the columns of P_1 are linearly dependent. A similar procedure can be used to reduce the order if the rows of Q are linearly dependent.

When, A, B, C are constant

$$h(t, t_0) = Ce^{A(t-t_0)}BU(t - t_0) \tag{7.12}$$

$$= Ce^{At}.e^{-At_0}BU(t - t_0) \tag{7.13}$$

$$= P(t)Q(t_0)U(t - t_0) \tag{7.14}$$

The connection between the procedure just given and the reduction procedure in Chapter 2, Section 4, is then expressed by the following result. This shows that reduction by the earlier procedure can be carried out when and only when it can be carried out by the procedure of Theorem 7.1.

Theorem 7.2 The columns of Ce^{At} are linearly dependent if and only if rank $\begin{pmatrix} sI_n - A \\ -C \end{pmatrix}$ is less than n for some $s = s_0$. Similarly the rows of $e^{-At}B$ are linearly dependent if and only if $(sI_n - A \quad B)$ has rank less than n for some $s = s_1$.

Proof If $Ce^{At}v \equiv 0$ for some $v \neq 0$, then

$$\mathscr{L}\{Ce^{At}vU(t)\} = C(sI_n - A)^{-1}v \equiv 0 \tag{7.15}$$

which by Chapter 2, equation (1.8), gives

$$0 = Cv = CAv = \cdots = CA^{n-1}v \tag{7.16}$$

or

$$\begin{pmatrix} C \\ CA \\ \vdots \\ CA^{n-1} \end{pmatrix} v = 0 \tag{7.17}$$

As v is non-zero, this shows that the matrix in (7.17) has rank less than n, whence (Chapter 2, Section 6) $sI_n - A$, C are not relatively (right) prime and $\begin{pmatrix} sI_n - A \\ -C \end{pmatrix}$ has rank less than n for some $s = s_0$. This argument can be reversed, and $e^{-A_0t}B$ can be treated similarly.

We now return to the time-dependent system. Let P and Q be reduced as in Theorem 7.1 until no further reduction can be made, and call the resulting matrices P_0 and Q_0. Then we have the following preliminary result.

Theorem 7.3 If the n_0 columns of P_0 and the n_0 rows of Q_0 are linearly independent, there exist θ_0, θ_1 such that the $n_0 \times n_0$ matrices

$$M'(\theta_0, \theta_1) = \int_{\theta_0}^{\theta_1} P_0^{\mathrm{T}}(\tau)P_0(\tau)\, d\tau \tag{7.18}$$

$$W'(\theta_0, \theta_1) = \int_{\theta_0}^{\theta_1} Q_0(\tau)Q_0^{\mathrm{T}}(\tau)\, d\tau \tag{7.19}$$

are each non-singular.

Proof By definition, for any matrix R,

$$\int_{\alpha}^{\beta} R^{\mathrm{T}}(\tau)\, d\tau = \left[\int_{\alpha}^{\beta} R(\tau)\, d\tau\right]^{\mathrm{T}} \tag{7.20}$$

whence M' is symmetric. Also for any constant vector v,

$$v^{\mathrm{T}}M'(\alpha, \beta)v = \int_{\alpha}^{\beta} \|P_0(\tau)v\|^2\, d\tau \geqslant 0 \tag{7.21}$$

so that $M'(\alpha, \beta)$ is positive semidefinite for all $\alpha, \beta > \alpha$. Therefore $M'(\alpha, \beta)$ can be written $H^{\mathrm{T}}H$ [compare Exercise 5.3] and it follows that if $v^{\mathrm{T}}M'(\alpha, \beta)v = 0$, then

$$v^{\mathrm{T}}H^{\mathrm{T}}Hv = \|Hv\|^2 = 0 \tag{7.22}$$

and so

$$Hv = 0 \tag{7.23}$$

and

$$M'(\alpha, \beta)v = 0 \tag{7.24}$$

The converse is certainly true, so that $v^{\mathrm{T}}M'(\alpha, \beta)v = 0$ if and only if $M'(\alpha, \beta)v = 0$. That is, the vectors satisfying $v^{\mathrm{T}}M'(\alpha, \beta)v = 0$ are precisely those lying in the kernel (the null space) of $M'(\alpha, \beta)$.

If $\delta \leqslant \alpha, \gamma \geqslant \beta$, (7.18) shows that

$$v^{\mathrm{T}}M'(\gamma, \delta)v = v^{\mathrm{T}}M'(\gamma, \alpha)v + v^{\mathrm{T}}M'(\alpha, \beta)v + v^{\mathrm{T}}M'(\beta, \delta)v \tag{7.25}$$

and as in (7.21) each of the terms in the sum is non-negative. Consequently

$$\text{kernel } M'(\gamma, \delta) = \text{kernel } M'(\gamma, \alpha) \cap \text{kernel } M'(\alpha, \beta) \cap \text{kernel } M'(\beta, \delta) \tag{7.26}$$

and so

$$\text{kernel } M'(\gamma, \delta) \subseteq \text{kernel } M'(\alpha, \beta) \tag{7.27}$$

It follows that the nullity (the dimension of the kernel) of $M'(\alpha, \beta)$ is non-increasing as β increases and as α decreases. [Compare Exercise 5.4, noting as in Chapter 1, Section 4 that rank + nullity = n_0.]

Choose any α, and consider nullity $M'(\alpha, \beta)$ as β increases. This is a non-increasing function taking integral values, and there must exist some β_M such that

$$\text{nullity } M'(\alpha, \beta) = \text{nullity } M'(\alpha, \beta_M), \quad \text{all } \beta > \beta_M \tag{7.28}$$

By (7.26) this means that

$$\text{kernel } M'(\alpha, \beta) = \text{kernel } M'(\alpha, \beta_M), \quad \text{all } \beta > \beta_M \tag{7.29}$$

Keeping β at β_M, now decrease α. There exists an α_M such that

$$\text{nullity } M'(\alpha, \beta_M) = \text{nullity } M'(\alpha_M, \beta_M), \quad \text{all } \alpha < \alpha_M \tag{7.30}$$

Moreover, by (7.26) and (7.29),

$$\text{kernel } M'(\alpha, \beta) = \text{kernel } M'(\alpha_M, \beta_M), \quad \text{all } \alpha < \alpha_M, \beta > \beta_M \tag{7.31}$$

In the same way we can find α_W, β_W such that

$$\text{kernel } W'(\alpha, \beta) = \text{kernel } W'(\alpha_W, \beta_W), \quad \text{all } \alpha < \alpha_W, \beta > \beta_W \tag{7.32}$$

Then we define

$$\theta_0 = \min(\alpha_M, \alpha_W), \qquad \theta_1 = \max(\beta_M, \beta_W) \tag{7.33}$$

We now show that nullity $M'(\theta_0, \theta_1) = 0$. Suppose to the contrary that $M'(\theta_0, \theta_1)$ is singular, and choose $v \neq 0$ such that $M'(\theta_0, \theta_1)v = 0$. Then for any $\alpha < \theta_0, \beta > \theta_1$, (7.31) shows that

$$v^{\mathrm{T}} M'(\alpha, \beta) v = \int_{\alpha}^{\beta} \|P_0(\tau)v\|^2 \, d\tau = 0 \tag{7.34}$$

which because $P_0(t)$ is continuous for all t implies

$$P_0(t)v = 0, \quad \text{all } t \tag{7.35}$$

contradicting the linear independence of the columns of P_0. The result for $W'(\theta_0, \theta_1)$ is established similarly.

Corollary The system is controllable (p.s.), and also observable, on $[\theta_0, \theta_1]$.

Proof We have, by (3.14) and (2.9),

$$M'(\theta_0, \theta_1) = \Psi^{\mathrm{T}}(\theta_0, t_1) M(\theta_0, \theta_1) \Psi(\theta_0, t_1) \tag{7.36}$$

where t_1 is some fixed time, $\Psi(\theta_0, t_1)$ is non-singular, and $M(\theta_0, \theta_1)$ is the controllability matrix. Similar results apply to $W'(\theta_0, \theta_1)$.

EXERCISE 7.1 Show that θ_0 and θ_1 in Theorem 7.3 depend in general on the initial choice of α preceding (7.28). Illustrate by means of a time-invariant system.

Now suppose that a system (7.1), (7.2) is given, and that $h(t, t_0)$ has been found and expressed as in (7.3). Then let $P(t)$ and $Q(t_0)$ have been reduced as in Theorem 7.1 until PQ has been expressed by

$$P(t)Q(t_0) = P_0(t)Q_0(t_0) \tag{7.37}$$

in which the n_0 columns of P_0 and the n_0 rows of Q_0 are linearly independent. We show that any system giving rise to the weighting pattern P_0Q_0 has order $n \geqslant n_0$.

Theorem 7.4 Suppose that the weighting pattern P_0Q_0 has $P_0(t)$, $Q_0(t_0)$ continuous for all time t, t_0, with the n_0 columns of P_0 and the n_0 rows of Q_0 linearly independent. If there exists a second decomposition $P_0(t)Q_0(t_0) = P(t)Q(t_0)$, with $P(t)$, $Q(t_0)$ continuous for all finite t, t_0 and respectively of dimension $m \times n$, $n \times l$, then $n \geqslant n_0$.

Proof We have for all t, t_0

$$P(t)Q(t_0) = P_0(t)Q_0(t_0) \tag{7.38}$$

and so

$$\left\{\int_{\theta_0}^{\theta_1} P_0^{\mathrm{T}}(\tau)P(\tau)\,d\tau\right\}Q(t_0) = \left\{\int_{\theta_0}^{\theta_1} P_0^{\mathrm{T}}(\tau)P_0(\tau)\,d\tau\right\}Q_0(t_0) \tag{7.39}$$

where θ_0, θ_1 are chosen as in Theorem 7.3 Hence

$$Q_0(t_0) = \left\{(M')^{-1}(\theta_0, \theta_1)\int_{\theta_0}^{\theta_1} P_0^{\mathrm{T}}(\tau)P(\tau)\,d\tau\right\}Q(t_0) \tag{7.40}$$

$$= RQ(t_0) \tag{7.41}$$

where R is a real constant $n_0 \times n$ matrix. [In the same way (though it is not required for the proof)

$$P_0(t) = P(t)T \tag{7.42}$$

where

$$T = \left\{\int_{\theta_0}^{\theta_1} Q(\tau)Q_0^{\mathrm{T}}(\tau)\,d\tau\right\}(W')^{-1}(\theta_0, \theta_1) \tag{7.43}$$

is a real constant $n \times n_0$ matrix.] But there exists no non-zero n_0-vector such that $v^T Q_0(t_0) = 0$ for all t_0, which implies that there is no $v \neq 0$ such that $v^T R = 0$. Hence $n \geqslant n_0$.

Corollary All decompositions (7.37), in which the columns of P_0 and the rows of Q_0 are linearly independent, have the same order n_0, which we shall call the *least order* corresponding to the weighting pattern (not the least order of the system).

Proof Let there be two decompositions PQ and $P_0 Q_0$ of h, in which P is $m \times n$, Q is $n \times l$, P_0 is $m \times n_0$, Q_0 is $n_0 \times l$; and the columns of P, P_0 and the rows of Q, Q_0 are linearly independent. Then, as in the theorem, $n \geqslant n_0$, and also $n_0 \geqslant n$, whence $n = n_0$.

EXERCISE 7.2 The conclusion of Theorem 7.4 was drawn from equation (7.41) and the linear independence of the rows of Q_0. Show where the assumption that the columns of P_0 are linearly independent enters into this conclusion. [In the non-singularity of $M'(\theta_0, \theta_1)$.]

EXERCISE 7.3 Let A in (7.1) be constant, let $B(t)$ be zero outside the interval [2, 3], and let $C(t)$ be zero outside the interval [0, 1]. Show that $h(t, t_0) \equiv 0$, which can be generated from a system of zero order. Show on the other hand that $B(t)$, $C(t)$ can be chosen so that the columns of P and the rows of Q are linearly independent. [Construct an example with $n = 1$.]

The preceding exercise illustrates the way in which the original question about least order has been weakened. As in the corollary of Theorem 7.3, the system is controllable (p.s.), and also observable, on [0, 3]. However it is not controllable (p.s.) on [0, 2], nor is it observable on the subsequent interval [2, 3]. This indicates how the requirement of causality must be introduced if the idea of least order is to be sharpened. When this is done, however, the strengthened reduction procedure may not lead to a system of the type (7.1), (7.2). These difficulties do not arise with constant linear systems, for which Theorem 7.2 shows that the least order corresponding to the weighting pattern is the least order of the system.

8 System similarity

We now establish a generalization, for time-dependent systems, of the corollary of Theorem 3.1 in Chapter 3.

Theorem 8.1 Let two systems have operational system matrices

$$P(p, t) = \begin{pmatrix} pI_n - A(t) & B(t) \\ -C(t) & 0 \end{pmatrix} \tag{8.1}$$

and

$$P_0(p, t) = \begin{pmatrix} pI_n - A_0(t) & B_0(t) \\ -C_0(t) & 0 \end{pmatrix} \tag{8.2}$$

in which A, B, C, A_0, B_0, C_0 are continuous for all t, and P, P_0 both have least order as defined in the corollary of Theorem 7.4. Then P, P_0 give rise to the same weighting pattern if and only if there exists a matrix $H(t)$ which for all finite t is non-singular and has continuous first derivative, and

$$P_0(p, t) = \begin{pmatrix} H^{-1}(t) & 0 \\ 0 & I_m \end{pmatrix} P(p, t) \begin{pmatrix} H(t) & 0 \\ 0 & I_l \end{pmatrix} \tag{8.3}$$

Proof Sufficiency follows as in the proof of Theorem 4.1. Let the two systems have transition matrices $\Psi(t, t_0)$ and $\Psi_0(t, t_0)$, and write for some fixed t_1,

$$C(t)\Psi(t, t_1)\Psi(t_1, t_0)B(t_0) = P(t)Q(t_0) \tag{8.4}$$

$$C_0(t)\Psi_0(t, t_1)\Psi_0(t_1, t_0)B_0(t_0) = P_0(t)Q_0(t_0) \tag{8.5}$$

As in (7.41), (7.42),

$$Q_0(t_0) = RQ(t_0) \tag{8.6}$$

$$P_0(t) = P(t)T \tag{8.7}$$

where R, T are real constant $n \times n$ matrices. From (8.7) and (7.40), (7.41) we obtain

$$\int_{\theta_0}^{\theta_1} P_0^{\mathrm{T}}(\tau)P_0(\tau) = \left\{\int_{\theta_0}^{\theta_1} P_0^{\mathrm{T}}(\tau)P(\tau)\, d\tau\right\}T \tag{8.8}$$

$$M'(\theta_0, \theta_1) = M'(\theta_0, \theta_1)RT \tag{8.9}$$

so that

$$R = T^{-1} \tag{8.10}$$

Then inserting P, Q, P_0, Q_0 from (8.4), (8.5) into (8.6), (8.7),

$$\Psi_0(t_1, t_0)B_0(t_0) = T^{-1}\Psi(t_1, t_0)B(t_0) \tag{8.11}$$

$$C_0(t)\Psi_0(t, t_1) = C(t)\Psi(t, t_1)T \tag{8.12}$$

On writing

$$H(t) = \Psi(t, t_1)T\Psi_0(t_1, t) \tag{8.13}$$

we find that (8.11), (8,12) become

$$B_0(t) = H^{-1}(t)B(t) \tag{8.14}$$

$$C_0(t) = C(t)H(t) \tag{8.15}$$

The matrix H in (8.13) is continuous and non-singular for all finite t, by the properties of the transition matrix, and has derivative

$$\dot{H}(t) = A(t)\Psi(t, t_1)T\Psi_0(t_1, t) + \Psi(t, t_1)T\left\{\frac{\partial}{\partial t}\Psi_0^{-1}(t, t_1)\right\} \tag{8.16}$$

$$= A(t)H(t) - H(t)A_0(t) \tag{8.17}$$

where we have used

$$\frac{\partial}{\partial t}\Psi_0^{-1}(t, t_1) = \Psi_0^{-1}(t, t_1)\left\{\frac{\partial}{\partial t}\Psi_0(t, t_1)\right\}\Psi_0^{-1}(t, t_1) \tag{8.18}$$

Equation (8.17) shows that $\dot{H}(t)$ is continuous for all finite t, and moreover gives

$$A_0(t) = H^{-1}(t)A(t)H(t) - H^{-1}(t)\dot{H}(t) \tag{8.19}$$

Comparison with equation (4.5) verifies 8.3.

The following theorem now allows us to regain the corollary of Theorem 3.1 in Chapter 3.

Theorem 8.2 If in (8.1), (8.2) the matrices A, B, C, A_0, B_0, C_0 are constant, then H is constant.

Proof Equation (8.13) gives

$$H(t_1) = T \tag{8.20}$$

and (8.14) then gives, for $t = t_1$,

$$B = TB_0 \tag{8.21}$$

which because B, B_0, T are constant matrices holds generally. Also, from (7.14), (8.6), and (8.10),

$$e^{-A_0t_0}B_0 = T^{-1}e^{-At_0}B \tag{8.22}$$

which with (8.21) gives

$$e^{-A_0t_0}B_0 = T^{-1}e^{-At_0}TB_0 \tag{8.23}$$

and so from the Taylor series

$$A_0^rB_0 = T^{-1}A^rTB_0 = (T^{-1}AT)^rB_0 \tag{8.24}$$

Then (8.24) gives

$$A_0^{r+1}B_0 = (T^{-1}AT)^{r+1}B_0 \tag{8.25}$$

$$= T^{-1}AT(T^{-1}AT)^rB_0 \tag{8.26}$$

$$= T^{-1}ATA_0^rB_0 \tag{8.27}$$

and the equations for $r = 0, 1, \ldots, n-1$ can be written

$$A_0(B_0 \quad A_0B_0 \quad \ldots \quad A_0^{n-1}B_0) = T^{-1}AT(B_0 \quad A_0B_0 \quad \ldots \quad A_0^{n-1}B_0) \qquad (8.28)$$

Now the system (8.2) has least order in the sense of the corollary of Theorem 7.4, so that the rows of $Q_0(t_0) = e^{-A_0t_0}B_0$ are linearly independent. By Theorem 7.2 and Chapter 2, Theorem 6.2, this implies that the matrix $(B_0 \; A_0B_0 \ldots A_0^{n-1}B_0)$ has rank n. Equation (8.28) therefore has the unique solution

$$A_0 = T^{-1}AT \qquad (8.29)$$

Inserting this in (8.13) we obtain

$$\begin{aligned} H(t) &= e^{A(t-t_1)}Te^{T^{-1}AT(t_1-t)} \\ &= T \end{aligned} \qquad (8.30)$$

which is constant.

On using Theorem 3.3 of Chapter 2, we may derive Theorem 3.1 of Chapter 3 from its corollary. This gives an alternative way of developing the material in Chapters 2 and 3.

Problems

1 If A and B are continuous functions of t, show that the controllability matrix $W(t_0, t_1)$ in equation (5.1) satisfies

$$\frac{\partial}{\partial t} W(t, t_1) = A(t)W(t, t_1) + W(t, t_1)A^{\mathrm{T}}(t) - B(t)B^{\mathrm{T}}(t); \qquad W(t_1, t_1) = 0$$

2 If A is continuous and $B \in D_R$, show that the controllability matrix satisfies

$$W(t_0, t_1) = W(t_0, t) + \Psi(t_0, t)W(t, t_1)\Psi^{\mathrm{T}}(t_0, t)$$

3 Find results for the observability matrix $M(t_0, t_1)$ analogous to those given for $W(t_0, t_1)$ in Problems 1 and 2.

4 Show that

$$\begin{pmatrix} \cos t_1 & -\sin t_1 \\ \sin t_1 & \cos t_1 \end{pmatrix} \begin{pmatrix} \cos t_2 & -\sin t_2 \\ \sin t_2 & \cos t_2 \end{pmatrix} = \begin{pmatrix} \cos(t_1+t_2) & -\sin(t_1+t_2) \\ \sin(t_1+t_2) & \cos(t_1+t_2) \end{pmatrix}$$

Interpret this as an example of equation (2.9).

5 Let the open-loop system S_0:

$$\begin{aligned} \dot{x}(t) &= A(t)x(t) + B(t)u(t) \\ y(t) &= C(t)x(t) \end{aligned}$$

be transformed by feed-back to the closed-loop system S_c:

$$\begin{aligned} \dot{x}(t) &= [A(t) - B(t)C(t)]x(t) + B(t)u(t) \\ y(t) &= C(t)x(t) \end{aligned}$$

Show that if S_0 is controllable (p.s.) on $[t_0, t_1]$, then so is S_c. [If u_0 drives x in S_0 from c_0 to c_1, along x_0, then $u_0 + Cx_0$ drives x in S_c from c_0 to c_1.]

6 Show that if S_0 in Problem 5 is observable on $[t_0, t_1]$, then so is S_c.

7 Show that in Problem 5, the controllability matrix for S_0 has the same range as the controllability matrix for S_c on any interval $[t_0, t_1]$, and similarly for the observability matrix. [Use methods similar to that suggested in Problem 5. Attempt also to find a direct proof.]

7 Notes and references

In accordance with the purposes of this book, references and attributions have generally been omitted in the text. The following notes will enable the reader to appreciate the outlines of the way in which the subject has developed. Further details will be found in the references.

Chapter 2

The early development of control theory was concerned with single-loop systems which were linear and constant. The design methods were based on frequency response and the Laplace transform. After 1955 the demands of rocket guidance caused attention to be transferred to time-varying and nonlinear systems, and to optimization. Here state-space methods and the calculus of variations are the appropriate mathematical tools. State-space methods were introduced notably by Bellman, Glicksberg and Gross (1958) and by Kalman (1960).

State-space methods have been applied more recently to industrial problems in an attempt to solve multivariable, constant, linear control problems. Much of Chapters 2 to 5 is concerned with bringing state-space and frequency-response methods into a single framework for this purpose (Rosenbrock, 1966).

Section 1.1 For the development of system matrices see Rosenbrock (1967a, b).

Section 3 The transformation of system similarity was extensively investigated [under the name *strict* (*algebraic*) *equivalence*] by Kalman (1963). Strict system equivalence is related on the one hand to transformations of differential equations (Gantmacher, 1959, p. 137, and also Section 2.1 of Chapter 2 in this book) and on the other to Howitt transformations (Newcomb, 1966, p. 328). System equivalence is similarly related to Howitt transformations. Theorem 3.3 (Rosenbrock, 1967b) provides an essential link between system similarity and strict system equivalence.

Section 5 See Rosenbrock (1970).

Section 6 The importance of Theorem 6.2(iii) was first emphasized and systematically developed by Kalman (1960, 1963) as a condition for controllability (p.s.). The result has an algebraic importance independent of its connection with controllability (p.s.). See also Rosenbrock (1967a) and Barnett (1970).

Section 7 The basic result in Theorem 7.1 was obtained, using vector-space methods, by Kalman (1962). For Algorithm 7.1 see Rosenbrock (1968). An alternative algorithm is given by Mayne (1968).

Section 8 For Exercises 8.2 to 8.4 and interesting further results see Barnett (1970).

Chapter 3

Section 1 The derivation of standard forms in Sections 1.1 and 1.2 is slightly adapted from Luenberger (1967). A less general treatment than Luenberger's had been given previously by Langenhop (1964). For the connection with minimal indices see Rosenbrock and Rowe (1970).

Section 2 Much of the difficulty in state-space methods lies in the absence of a convenient (and unique) standard form under system similarity. This deficiency is supplied here by developing a standard form under strict system equivalence and relying on Theorem 3.3 of Chapter 2.

Section 3 The corollary of Theorem 3.1 is a basic result which was proved by Ho and Kalman (1966) and by Youla and Tissi (1966). The corresponding theorem for time-dependent systems had previously been announced without proof by Kalman (1963), but requires some qualification (Kalman, Falb and Arbib, 1969, p. 319).

Section 4 Theorems 4.1 and 4.2 show that the system matrix is more than a formal device.

Section 5 Compare Gilbert (1963), Kalman (1963), Ho and Kalman (1966), and Youla and Tissi (1966).

Section 6 Compare Gilbert (1963). The standard form given in Section 2 of this chapter permits the relatively simple and complete treatment given here (Rosenbrock, 1970).

Section 7 The basic work on this subject is due to McMillan (1952). The treatment given here is slightly adapted and generalized.

Problems For Problems 1 to 7 see Rosenbrock (1967c). For Problem 11 see Hannan (1969).

Chapter 4

Section 2 The treatment is adapted from Kalman, Ho and Narendra (1963).

Section 3 The algorithm leading to (3.15) is new. That leading to (3.35) was given by Ho and Kalman (1966).

Chapter 5

Sections 2, 3 See the note on Chapter 2, Section 6. Compare recent work by Basile and Marro (1969, 1970). The definition in Section 2.4 is due to Simon and Mitter (1968). Section 2.5 is based on Brockett and Mesarovic (1965). For Section 2.6 see Rosenbrock (1966). The basic work on observers, Section 3.1, was done by Luenberger (1964, 1966), and the treatment here follows Newmann (1969).

Section 4 Special forms of Theorem 4.2, Corollary 2 were well known, but the first general proof for real A, B, C seems to be Wonham's (1967). Compare Popov (1964), Luenberger (1964, 1966), Simon and Mitter (1968). Minimal indices were introduced in Rosenbrock and Rowe (1970).

Section 5 The results here were obtained, using other methods and in slightly different form, by Luenberger (1964, 1966).

Sections 6, 7 See Rosenbrock (1969). These results are immediately useful for design purposes. Theorems 6.5 and 6.6 are due to Ostrowski (1952).

Problems For Problems 1 to 5 see MacFarlane and Rosenbrock (1970).

Chapter 6

This chapter leans heavily on the papers by Kalman, Ho and Narendra (1963) and Youla (1966).

Sections 1–3 This material is largely standard. The important Theorem 3.2 is due to Kalman (1963); see also Youla (1966).

Section 4 The extent to which operational system matrices can be pursued is completely unexplored.

Sections 5, 6 The definitions here, which are not standard, are aimed at the needs of applications rather than those of abstract system theory. The treatment is adapted from Kalman, Ho and Narendra (1963) and has benefited from Brockett (1970).

Sections 7, 8 These follow the paper by Youla (1966). Theorem 7.3 gives an alternative approach to Youla's Lemma 3; compare Weiss (1968). Theorems 7.4 and 8.1 are Youla's proof of a weakened form of the theorem announced by Kalman (1963). Theorem 8.2 again follows Youla (1966), whose paper contains more general results than those quoted here.

General references

The background mathematics will be found in Rosenbrock and Storey (1970). Zadeh and Desoer (1963) give an account of the subject before the recent period of great activity began. A good introduction to state-space methods is given by Brockett (1970). Kalman, Falb and Arbib (1969) give an abstract treatment of part of the material covered here: a useful short account will be found in Arbib and Zeiger (1969).

Future work

The most obvious gap in Chapters 2 and 3 is that necessary and sufficient conditions for strict system equivalence are not known. It would be more satisfactory, if such conditions were known, to treat systems of least order as a special case of the general theory. The answer to such questions probably lies in the theory of pencils of matrices.

Work is also needed on practical design methods for multivariable control systems. Sections 6 and 7 of Chapter 5 give one approach. Another, based on Corollary 2 of Theorem 4.2 in Chapter 5, has been suggested by Luenberger (1966) and Simon and Mitter (1968).

References

ARBIB, M. A. and H. P. ZEIGER, 1969, On the relevance of abstract algebra to control theory, *Automatica*, **5**, 589–606.

BARNETT, S., 1970, Matrices in control theory and other applications (to be published), Van Nostrand Reinhold.

BASILE, G. and G. MARRO, 1969, On the observability of linear time-invariant systems with unknown inputs, *J. Optimization Theory and Applications*, **3**, 410–415.

BASILE, G. and G. MARRO, 1970, On the perfect output controllability of linear dynamic systems, to be published.

Bellman, R. E., I. Glicksberg and O. A. Gross, 1958, Some aspects of the mathematical theory of control processes, *R.A.N.D. Report*, R-313.
Brockett, R. W. and M. D. Mesarovic, 1965, The reproducibility of multivariable systems, *Math. Analysis and Applications*, **1**, 548–563.
Brockett, R. W., 1970, Finite dimensional linear systems, John Wiley.
Gantmacher, F. R., 1959, The theory of matrices, vols. 1, 2, Chelsea Publishing Co.
Gilbert, E. G., 1963, Controllability and observability in multivariable control systems, *J. S.I.A.M. Control*, ser A, **2**, 128–151.
Hannan, E. J., 1969, The identification of vector mixed autoregressive-moving average systems. *Biometrika*, **56**, 223–225.
Ho, B. L. and R. E. Kalman, 1966, Effective construction of linear state-variable models from input/output functions, *Regelungstechnik*, **14**, 545–548.
Kalman, R. E., 1960, On the general theory of control systems, *Proc. First I.F.A.C. Congress*, **1**, 481–491, Butterworth.
Kalman, R. E., 1962, Canonical structure of linear dynamical systems, *Proc. Nat. Acad. Sci.*, **48**, pp. 596–600.
Kalman, R. E., 1963, Mathematical description of linear dynamical systems, *J. S.I.A.M. Control*, ser A, **1**, 152–192.
Kalman, R. E., 1964, On canonical realisations, *Proc. Second Allerton Conference on Circuit and System Theory*, 32–41.
Kalman, R. E., 1966, On structural properties of linear, constant, multivariable systems, *Third IFAC Congress*, paper 6A.
Kalman, R. E., P. L. Falb and M. A. Arbib, 1969, Topics in mathematical system theory, McGraw-Hill.
Kalman, R. E., Y. C. Ho and K. S. Narendra, 1963, Controllability of linear dynamical systems, *Contributions to the theory of differential equations*, vol. 1, pp. 189–213, Interscience.
Langenhop, C. E., 1964, On the stabilization of linear systems, *Proc. Am. Math. Soc.*, **15**, 735–742.
Luenberger, D. G., 1964, Observing the state of a linear system, *I.E.E.E. Trans.*, **MIL-8**, 74–80.
Luenberger, D. G., 1966, Observers for multivariable systems, *I.E.E.E. Trans.*, **AC-11**, 190–197.
Luenberger, D. G., 1967, Canonical forms for linear multivariable systems, *I.E.E.E. Trans.*, **AC-12**, 290–293.
MacFarlane, A. G. J. and H. H. Rosenbrock, 1970, New vector-space structure for dynamical systems, *Electronics Letters*, **6**, 162–163.
McMillan, B., 1952, Introduction to formal realizability theory, *Bell. System Tech. J.*, **31**, 217–279, 541–600.
Mayne, D. Q., 1968, Computational procedure for the minimal realization of transfer-function matrices, *Proc. I.E.E.*, **115**, 1363–1368.
Newcomb, R. W., 1966, Linear multiport synthesis, McGraw-Hill.
Newmann, M. M., 1969, Design algorithms for minimal-order Luenberger observers, *Electronics Letters*, **5**, 390–392.
Ostrowski, A. M., 1952, Note on bounds for determinants with dominant principal diagonal, *Proc. Am. Math. Soc.*, **3**, 26–30.
Rosenbrock, H. H., 1966, On the design of linear multivariable control systems, *Proc. Third I.F.A.C. Congress*, paper 1A.
Rosenbrock, H. H., 1967a, Transformation of linear constant system equations. *Proc. I.E.E.*, **114**, 541–544.

ROSENBROCK, H. H., 1967b, On linear system theory, *Proc. I.E.E.*, **114**, 1353–1359.

ROSENBROCK, H. H., 1967c, Efficient computation of least order for a given transfer function, *Electronics Letters*, **3**, 413–414.

ROSENBROCK, H. H., 1968, Computation of minimal representations of a rational transfer-function matrix, *Proc. I.E.E.*, **115**, 325–327.

ROSENBROCK, H. H., 1969, Design of multivariable control systems using the inverse Nyquist array, *Proc. I.E.E.*, **116**, 1929–1936.

ROSENBROCK, H. H., 1970, Properties of linear constant systems, *Proc. I.E.E.*, **117**, 1717–1720.

ROSENBROCK, H. H. and A. ROWE, 1970, Allocation of poles and zeros, *Proc. I.E.E.*, to be published.

ROSENBROCK, H. H. and C. STOREY, 1970, Mathematics of dynamical systems, Nelson.

SIMON, J. D. and S. K. MITTER, 1968, A theory of modal control, *Information and Control*, **13**, 316–353.

WEISS, L., 1968, On the structure theory of linear differential systems, *J. S.I.A.M. Control*, **6**, 659–680.

WONHAM, W. M., 1967, On pole assignment in multi-input controllable linear systems, *I.E.E.E. Trans.* **AC-12**, 660–665.

YOULA, D. C., 1966, The synthesis of linear dynamical systems from prescribed weighting patterns, *J. S.I.A.M.*, **14**, pp. 527–549.

YOULA, D. C. and P. TISSI, 1966, *n*-Port synthesis via reactance extraction, Part I, *I.E.E.E. International Convention Record.*

ZADEH, L. and C. A. DESOER, 1963, Linear system theory, McGraw-Hill.

Subject index

Section numbers are shown in parentheses